KU-095-919

Praise for *Our Wild Farming Life*

'I raced through this beautiful story with mounting awe and excitement. What Lynn and Sandra have achieved on their croft in Scotland is a miracle of rebirth on land that most would have considered beyond hope. Their success is testament to the wisdom of working with nature rather than battling against it and their pragmatism, honesty and openness to new and old ideas shines through on every page. I hope it inspires legions of new farmers.'

ISABELLA TREE, author of *Wilding*

'A ripping good account of the guts, luck and perseverance it takes to create a productive and healthy farm or croft that jumps the rails of our conventional industrial agriculture.'

NICK OFFERMAN

'Full of refreshing honesty and a burning passion to reconnect food, communities and nature, what Lynn and Sandra have achieved is nothing short of incredible. These two are the real deal; humble pioneers during a critical time working selflessly to empower others. Lynbreck Croft is the embodiment of how humans should approach food production, and this book serves as an invaluable resource for anybody who has an interest in a regenerative future.'

HUW RICHARDS, author of *The Vegetable Grower's Handbook*

'The inspiring story of two courageous women who took the leap and embraced a whole new way of life. Lynn and Sandra, I salute you!'

KATE HUMBLE, broadcaster and author of *Home Cooked*

'This is a story that needed to be told. I defy anyone to be in the presence of Lynn and Sandra for ten minutes and not come away inspired and with a smile.

'I know I did.

'Arriving in the Highlands with a dream and very little else apart from determination, enthusiasm, passion, warmth and humility.

'Not afraid to learn from mistakes and to seek help and advice.

'The Highlands can be a welcoming place, but you have to earn it. Folk here won't beat a path to your door unless you leave it wide open.

'And they did it. Neighbours embraced their dream and wanted to make it happen. To be a part of it. The gift of a cow. The gift of fuel. The gift of thousands of years of experience, knowledge and a love for the land.'

EUAN MCILWRAITH, BBC TV and Radio presenter

'Many of us dream of going "back" to the land. Lynn Cassells and Sandra Baer have made it happen. In this inspirational, honest and quietly revelatory book they show how farming is not a lifestyle fantasy but a way of living, working and eating that is in partnership – with the land, the community, animals and each other. This is not a step back but the key to all our futures.'

PATRICK BARKHAM, author of *Wild Child*

OUR WILD FARMING LIFE

OUR WILD FARMING LIFE

Adventures on a Scottish Highland Croft

LYNN CASSELLS AND
SANDRA BAER

Chelsea Green Publishing
White River Junction, Vermont
London, UK

Commissioning Editor: Jonathan Rae
Project Manager: Patricia Stone
Developmental Editor: Muna Reyal
Copy Editor: Susan Pegg
Proofreader: Nikki Sinclair
Designer: Melissa Jacobson

Printed and bound in Great Britain by Clays Ltd, Elcograf S.p.A.
First printing February 2022.
10 9 8 7 6 5 4 3 2 22 23 24 25 26

ISBN 978-1-64502-070-7 (hardcover) | ISBN 978-1-64502-165-0 (paperback) |
ISBN 978-1-64502-071-4 (ebook) | ISBN 978-1-64502-072-1 (audio book)

Library of Congress Cataloging-in-Publication Data
Names: Cassells, Lynn, 1978- author. | Baer, Sandra, 1983- author.
Title: Our wild farming life : adventures on a Scottish highland croft /Lynn Cassells and Sandra Baer.
Description: White River Junction, Vermont : Chelsea Green Publishing,[2022]
Identifiers: LCCN 2021056924 (print) | LCCN 2021056925 (ebook) | ISBN
9781645021650 (paperback) | ISBN 9781645020707 (hardcover) | ISBN 9781645020714 (ebook)
Subjects: LCSH: Cassells, Lynn, 1978- | Baer, Sandra, 1983- | Farm
Life—Scotland—Highlands. | Farmers—Scotland—Highlands—Biography. |
Crofts—Scotland—Highlands. | Farms, Small—Scotland.
Classification: LCC S522.G7 C37 2022 (print) | LCC S522.G7 (ebook) | DDC
630.9411/5—dc23/eng/20211202
LC record available at https://lccn.loc.gov/2021056924
LC ebook record available at https://lccn.loc.gov/2021056925

Chelsea Green Publishing
85 North Main Street, Suite 120
White River Junction, Vermont USA

Somerset House
London, UK

www.chelseagreen.com

To Elaine, Ueli, Brian, Maree and Clare

Contents

Introduction

We never meant to be farmers.

It was a warm, sunny late-summer's day as we drove down the bumpy stone track and had our first experience of the Lynbreck view. The vast expanse of the heather-carpeted valley floor merging into the Scots pines of Abernethy Forest, then slowly climbing the lower slopes of the granite hills behind, with Cairn Gorm – the mountain that lends its name to the whole range – taking centre stage, was something we'd never forget.

The sales brochure had described Lynbreck Croft as: 'A rare opportunity to purchase an attractive registered croft located within the Cairngorms National Park, extending to 150 acres and enjoying a spectacular setting … with its mixed topography and stunning views to the south into the heart of the Cairngorm mountains, there is a great deal of potential to further develop the property and enhance its amenity, agricultural and woodland appeal.'

Stunning doesn't even come close to describing the view. On that day in August, there was a slight haze in the air, making the expanse before us gently vibrate. The haze gave everything a soft edge and it felt as if we were looking at a watercolour painting or chalk drawing.

We were met by the owner and, after a quick tour of the small wooden cabin that comprised the living space, we set off to explore the land. After an hour or so of wandering through fields and woodland, we finally stopped on the side of a hill, collapsing into a springy mat of flowering purple heather and taking in deep breaths of Highland air, which filled our nostrils and lungs with the intense floral aroma that was all around us. Here, we realised that this was the land where we wanted our life story to unfold.

I was lucky to grow up on the edge of a medium-sized town in Northern Ireland, where I was out from dawn to dusk, playing with friends, exploring the local countryside, building dens and racing down the hill at the front of our house as fast as I could on my skateboard.

Looking back, it was a wild and free existence, and I would delve into my own imagination to create daily adventures that involved flying into space with my sister from our coal bunker or fighting off an army of invaders from my friend Timothy's tree house.

As I matured into my teenage years in the nineties, I experienced some of the worst of 'the Troubles', living in a part of the province that was called the 'murder triangle'. I had friends who had lost parents in a spate of what we called 'tit for tat' killings that went on for years between warring Irish nationalists and British unionists. It was normal to be in bomb scares or to hear bombs going off: the sound of a deep, heavy reverberating thud that would sometimes make the ground shake as the violent tremor vibrated in ripples through the air.

When you grow up in that culture and it's all you know, it is a strange normality. Protestants lived on one side of the town and Catholics on the other, divided only by an invisible line, which was nevertheless clearly plotted in the minds of every member of the local community. But, as I grew older, I began to realise that this was anything but normal and soon, something happened to make me want to leave this country for good.

Introduction

The summer of 1996 was one of the worst prolonged periods of intense unrest in the history of the Troubles, with the flashpoint just a few miles from where we lived. I remember sitting in my bedroom, listening to the nearby rioting and gunfire as I watched a spent police flare drift eerily down from the sky on a little white ghostly parachute and land gently in our back garden. I felt real fear that night. A few months later, I went to a pub in Belfast with a friend on a Friday night and a gunman burst through the doors and shot an off-duty policeman at the bar. I made eye contact that night with a masked killer, and my decision to leave this country was made.

A month after I turned nineteen, I moved to Birmingham to read archaeology at the University of Birmingham and would go on to complete my master's in archaeology over ten years later at University College London. In between my studies, I had various jobs – everything from archaeologist to teacher, ski resort host to youth worker, all the while trying to find what it was that I wanted to do with my life. I could never settle for long as a sense of boredom and feeling of futility would creep into the cracks of whatever I was doing at that time.

I started to volunteer with the National Trust, a large conservation charity, and learned about the apprenticeship scheme they ran for people like me, who wanted to retrain and work with nature in the outdoors. The post was for a ranger; a practical, hands-on job that would involve lots of conservation work in a range of habitats, balanced with some engagement with the general public who visited the many hundreds of landholdings the National Trust owns. I spent the next eighteen months gaining experience and preparing my application with such focus and drive that nothing was going to get in my way, until, finally, the day came when I could call myself an apprentice ranger and the great outdoors became my new office.

Sandra grew up in a small town near Zürich in the north of Switzerland with her parents and younger brother, living in a traditional Swiss country house. Sandra's dad was born there and over the years had watched how the little farm village of Kloten expanded rapidly after the construction of the nearby international airport. Even though the family house is now surrounded by sprawling housing estates and the green fields with their summer song of chirring crickets have become few and far between, Sandra got to live a childhood that was filled to the brim with nature and the outdoors.

Sandra and her brother were allowed an even more feral existence during their annual holidays in their mother's home country of Scotland when, during camping trips to Sutherland and stays on a farm in Fife, they were given free rein to explore the countryside to their hearts' content. Looking back, it's clear to see that those experiences shaped her life choices, setting her on a path that would eventually lead to a career in the natural world.

The Swiss education system offers an in-depth apprenticeship programme for a wide range of professions and practical experiences, and being hands-on appealed to Sandra rather than academia. However, at just fifteen, the young age at which apprenticeship training starts, she lacked the confidence to dive into the rather male-dominated world of her first-choice career path of forestry and opted for a more conventional training in the Central Library in Zürich. While she found the access to millions of books fascinating, after several years of working in an entirely indoor environment, the allure of travel and adventure became too hard to resist. She packed her bags and headed overseas to Canada to work on a ranch in British Columbia before training horses in Alberta, which gave her a whole new outlook on life in a country she had adopted as her second home.

Sandra discovered a love for a way of living where the old ways merged with the modern world, where days were spent on horseback, herding cattle to new pastures, and where the pace was set by the animals and the landscape they found themselves in. It encouraged her to learn about herself and, for the first time, a quiet confidence was allowed to develop. Her ranch hosts' parting words of 'always follow your dreams' awoke a yearning for a way of living that was more connected to the land – and that could no longer be quelled.

It was a gloriously sunny day in July 2012 when I pulled into the short-stay car park at Heathrow Airport. My work team had just recruited a new apprentice ranger and I was going to be their line manager, excited to put my own three years' experience into training up the next generation. We'd hired a woman from Switzerland, someone who had shone through at the interview, and I'd offered to help her find somewhere to live and show her the area before starting work a few months later.

Within 48 hours of being together, we realised the line between our new professional relationship and our personal one were becoming increasingly blurred. I've never felt as drawn to anyone in my life and the feeling was mutual. We parted at the airport with an understanding that something more powerful was happening between us. Before she left Switzerland for the last time, Sandra told her close friends that not only was she moving to a new country, a new house and a new job, but she would also be dating her new boss. Needless to say, they were intrigued to see how this would go.

Two months later on a wet September evening, we sat in stationary traffic in my little Volkswagen Polo on the motorway. The rain was lashing down with such ferocity that the windscreen wipers couldn't keep up. Combined with the glare of oncoming traffic, it had become virtually impossible to drive and the cars around us eventually slowed to a grinding halt. We'd just had a weekend in Wales visiting a friend and, as we sat with nowhere to go, the conversation turned to our individual hopes and dreams for the future. We each shared our love of being outdoors and of nature, passions that had led us both to completely change our careers to where we found ourselves now. But there was something else that ran deeper, a yearning for a wilder, freer way of living.

The more we talked that night, a clearer vision evolved between us of a life on the land, stemming from the deep desire within us both to reconnect with the earth beneath our feet. We grew even closer, sharing private feelings of frustration and unease about living in the world as it was presented to us today: the rush, the pressure, the traffic, the pollution, the rubbish and the obsession with material wealth. We felt increasingly adrift from what to us was the 'real world' – one where life sways in harmony with the natural environment – because that world had become obscured behind a thick screen of twenty-first-century smog.

The dream was to have time to grow our own food, to cook and bake with fresh eggs laid by our free-ranging hens, to gather and process our own firewood, to preserve and ferment our summer harvest, and to spend time appreciating and learning from the land around us. It was an idyll, a dream of a different way of living, but one where we knew bills would still have to be paid and ways found to make enough income to cover our annual costs. Even at this stage, we never imagined farming as an option, focusing instead on setting aside an acre or two of our imaginary plot for a small campsite or maybe even a few glamping rentals.

That winter, we spent so many evenings in front of the fire in our little cottage in the middle of the estate we were working on, gathering ideas of what this dream might look like. On an A4 page the words 'Big Idea' were written in the top right-hand corner and we started to brainstorm key thoughts that would help us to understand exactly what we were thinking of. Within minutes, the page was filled with words: bees, produce, goats, food smoker, hens, polytunnel, campsite, woodland, courses, shop, clay oven, communal fire pit, hedges, pigs, pond and lots more. We expanded onto other pages that were given titles: Online presence, Marketing and Planning, and so on, as we tried to turn our dreamy vision into a workable plan. The process of putting everything down on paper made it all feel a bit more real, that little bit closer within our reach. It was such a fun and exciting time of dreaming, when anything was possible and where our imaginations could run wild. A time that we look back on now with such happy memories.

I bought a book called *The Financial Times Guide to Business Start Up 2013*, spending evenings reading and making notes, trying to answer questions such as: What do you want? Why will you succeed? Why might you fail? What are your ideas? What is your market? This was a very different and new world to the one of nature and of practical outdoor work that we were currently immersed in. But, while it used jargon and phrases that were unfamiliar, there were a lot of basic principles that just seemed like common sense, such as 'make more money than you spend', 'be nice to your customers' and 'find and share your unique identity'.

Neither of us have ever been motivated by earning money for the accumulation of wealth to buy more material goods. It's always been a means to an end, to earn enough to pay the bills and allow us enough to live on. At the time, Sandra was starting out as an apprentice ranger on an annual salary of £12,000 a year and I had secured a permanent position as a ranger with my pay taking a jump up to £17,000. By the time rent on our cottage and bills were paid, living in one of the most expensive corners of the UK, there was little left to spend on anything else. While quality of life was what we were trying to achieve, we knew that trying to understand our possible future costs would be a crucial part of our planning. How much money did we have in savings? What was our maximum budget? What fees might we encounter like estate agents' and government taxes? The more we crunched the numbers, the more we built an idea of what we could afford. However, even with our combined resources, the high price of land in the area where we were currently living was prohibitively expensive for what we were after, and we began to accept that we would have to look much further afield.

And while evenings were taken up with dreaming, our work days involved long hours of physical labour with tasks varying from tree felling and fencing, to path building and strimming. Our 'office' would alternate from the middle of an ancient woodland surrounded by giant, gnarly old oak trees, chalk grasslands full of incredible wildflowers like bee orchids, harebell and field scabious, or waterside by the River Thames where cormorants would perch on trees, their wings outstretched to dry before their next fishing session.

We were immersed day to day in real-time ecology, developing a deep understanding of the intricacies of how nature works as well as how to identify different plants, mammals and insects. We developed our practical skills, learning how to drive a tractor, operate a chainsaw and hang gates, as well as our engagement skills by working with volunteers and delivering guided tours and short courses. And in the climate of the southeast of England, famously the warmest and driest in the UK, we enjoyed many long, hot summers and crisp, clear winters.

We absolutely loved the essence of what we did, the landscape we worked in and what it provided us with. But, despite the perceived idyll of our ranger careers, cracks were starting to appear as our team came under greater pressure to make money through product sales and public engagement events, taking us further away from the jobs we loved and the outdoor office we had become so used to. As the property we lived in and worked on became increasingly busier with visitors and our roles began to change, it made us seriously question our future, wondering if this might be the point to make our break. It was a confusing, unsettling and stressful time, as we went from valuing our jobs and home to feeling that perhaps it was a place that we no longer belonged.

A pivotal turning point arrived when we realised that only we could create the life of our dreams, not an employer. We're both headstrong, a trait which had led us to this point in our lives, and the idea of working for ourselves certainly appealed. But even at this stage, farming had still not really entered our minds as an option. Growing food? Yes. Diversifying income from the land? Yes. But farming? Not exactly.

I had been struggling particularly with my job, my patience wearing thin as I became increasingly vocal about things I simply couldn't agree with. I've never been very good at just keeping my mouth shut. I can do it to a point but, being impulsive to the core, my true feelings tend to erupt like a bottle of fizzy drink that has been opened after a vigorous shaking. I handed in my notice twice, withdrawing it both times as I got scared, willing to take any small promise of change as a reassurance that things would improve. Sandra, ever calm and supportive, rode this rocky road with me, herself becoming more and more frustrated as our desire to leave grew stronger. But could we really leave this all behind and, if so, where on earth would we go?

Due to her family connections, Sandra felt a strong draw to Scotland and I was open to moving anywhere. As an introduction for me, we took a holiday up to the Cairngorms National Park, a landscape of mountains, moorland and forest three hours north of the central belt of Glasgow and Edinburgh. We spent a couple of nights camping beneath a canopy of trees and stars, the weather behaving itself to showcase the land in its finest, sunny glory. We spent our days hiking, ascending steep slopes through ancient Caledonian woodland of Scots pine, rowan, birch and juniper that was slowly recolonising the bare hills above.

On the second evening, we settled into camp and warmed a dinner of tinned haggis on a little one-ring gas burner. Being in Scotland, it seemed like the appropriate meal to have and, in spite of not having a clue and not daring to think about what bits of sheep innards were in it, it was utterly delicious. Afterwards, we took a stroll, finding a bench on the edge of the forest and sitting silently, mesmerised by the scene in front of us, feeling humbled and slightly intimidated by the sheer power of the landscape at our feet. There lay a dense, squat juniper forest between us and the mighty Cairngorm range, a series of hills and mountains that are very different to places like the Alps or the Dolomites in central Europe with their jagged, pointy tops, which are in some cases up to four times as high as the Cairngorms. But the Cairngorms, which are part of a larger range known as the Grampians, are no less impressive, their rounded tops a sign of millennia of weathering, resting on the earth as some of the oldest mountains in the world. I remember saying to Sandra, 'Imagine living here. Imagine if that was your view.' Little did we know that behind us, just two miles north as the crow flies, sat Lynbreck Croft.

CHAPTER 1

A Leap of Faith

'I got it. I got the job.' I could barely make out the sentences as Sandra called with her news. The job in question was a six-month seasonal ranger post with the National Trust for Scotland on the Isle of Arran, a small island off the west coast of Ayrshire. We had decided that in order to seriously look for land in Scotland, we would need to be in Scotland as our base in England was nearly seven hours from the border, making any viewings very difficult and practically impossible. The position came with accommodation in Brodick Castle, a grand and impressive Scottish baronial castle owned by the charity on the edge of the largest village and main ferry port of Brodick. The small flat would be big enough for us both, with Sandra's wage covering our living costs.

I often look back to those times and try to remember how it felt. How do you make that leap? How do you know when the time is right? I remember feelings of trepidation and occasional surges of fear, the kind that rises up in your chest so quickly that you can hardly breathe, as we turned away from lifelong friends, from my family – including my only sister – and from my job that had prospects, a pension and many other perks. But this opportunity felt like our ticket to our dream. It was now or never. This was our chance.

I remember how completely exciting and liberating it felt, and the support from our friends was incredible. Sandra had an easy time in telling her very understanding and supportive boss of her ensuing departure who had apparently seen it coming for some time, whereas I had the slightly trickier job of telling mine that two of us would be leaving the team. I watched the initial shock on her face turn into a big smile, followed by the warmest hug, as she fully agreed that this was the best move for us, totally supporting our approach to give it a go. It felt as though we weren't just following our own dream, we were following the dreams of many around us who, deep down, also had the same hopes of following theirs one day.

Before long, it was time to fill our two little cars with the final bits and pieces, and drive north to a new country and a new life. I had found work with a small conservation charity planting trees a few hours away on the mainland and so, while weekdays were spent apart, weekends were packed full of adventures hiking, camping and exploring our way around Arran. After finishing her seasonal contract on the island, Sandra also found work tree planting with a small woodland management company, which gave her a valuable insight into running a small business, and we moved into a whitewashed mid-terrace cottage in the historic market town of Moffat, just a few miles to the north of the English border.

We allowed ourselves a couple of months to settle in and then our quest for land began on 1 January 2015. We had just two points on our 'must-have' list: anywhere in Scotland (being in the country was a good start) and a minimum size of around five acres (but this wasn't hard and fast, as a couple of acres more would be fine too). We didn't care if there was a house or buildings, focusing instead on the land and happy to live in a caravan, at least until we got set up.

In a short space of time, our simple search evolved into a military operation as we compiled detailed and extensive lists of every estate and land agent that we could find across the whole of Scotland. Once work was finished for the day, our evenings were spent going through the lists, starting to get an idea of what kind of properties were available as well as their costs and locations. We sat side by side on the couch, often without speaking, in focused silence. The first few months were frustratingly slow, with very little coming up that was of interest, making us doubt the way we were going about things. It was hard not to get downhearted or despondent, especially as we had literally come so far, but when either of us would have a low moment, the other would say, 'Be patient. We have to be patient. If it was this easy, everyone would be doing it.' And so the search continued until finally, in spring, we had our first breakthrough – or so we thought.

A plot had become available located to the north of Fort William on the west coast, with spectacular views of the Ben Nevis mountain range that boasts the highest peak in the UK. We could make it up there and back within a day, so we booked a viewing and the following weekend made the road trip north. It was a bigger plot than we had been wanting but, on paper, offered lots of potential. After a short wander, however, it soon became clear this would not be the place for us, the ground being mostly wet and boggy, and right next to a very busy tourist monument that attracted endless streams of coach tours packed with visitors from all over the world. It became the first of many disappointments, weekend after weekend, as our hunt intensified.

In those early days, I remember catching Sandra looking at a new property she had found, somewhere that had clearly caught her eye as; week after week, she kept revisiting the sales page.

Initially, her attention had been drawn by the utopian pictures that showed a small traditional farming homestead, with a few stone buildings and a wooden cabin, set against the backdrop of a large and impressive mountain range. It was a scene so perfect that it was hard to imagine it was a real place and up for sale just a few hours to the north of where we were living. Not only was the location ideal, but the description of the land that came with the chocolate-box homestead taunted and teased her. When she eventually decided to share it with me, feeling brave after an evening of fine food and red wine when spirits were high, I understood: the draw of the land intoxicating us both – and it wasn't just the effect of the booze.

But this was pointless. Yes, it was incredible, but it was way over our budget. We had to forget it. Our resolve to do so was genuine and well intentioned but ultimately futile. This place that Sandra had found grew into a secret, forbidden crush; one that you know is so wrong but that you want so badly. Instead of erasing it from our minds, it became all either of us could think about, but we never spoke of it, not wanting to get the other's hopes up when individually, it felt like an impossible dream.

'Well, we'll be driving right past it. Should we just arrange a quick look anyway?'

Quite a few months later, we had a busy weekend of viewings, starting in Aberdeenshire in north-east Scotland, followed by another one later that day just south of Inverness. As the route was finalised, we realised it would involve driving directly past the property we had secretly fallen in love with. Sandra suggested a visit would do no harm as I tried my best to casually agree with a forced air of nonchalance that spectacularly failed. We were going to see it, in real life, the excitement barely contained behind a false wall of casual exteriors that we erected as protection from the possibility of crushing disappointment.

The two-hour viewing of Lynbreck Croft felt like the shortest, most intense and utterly overwhelming blink of a moment in our lives. When we finally left, conscious not to overstay our welcome but desperately eking out every last second, it was clear what we were driving away from and what we had started our journey towards. It felt so exciting, but no sooner than the land was behind us, the reality of the next challenge was clear. We didn't have all the money to buy this place, so how would we ever make this happen?

After returning home, we immediately appointed a solicitor who specialised in crofting law to begin the initial process of officially noting an interest to purchase, a step which committed us to at least some financial expenditure. But the reality was that we were well short of what we needed. Getting a bank mortgage or loan would be impossible as moving to Lynbreck would involve leaving behind our paid employment. In a final act of desperation, we started to share our position with friends and family in the hope that they might help us come up with inspired solutions to our problem. Buying some land was all we had talked about for years, so it came as no surprise to them that our plans had finally started to progress. We had family that wanted to help but didn't have the available funds we needed in the short term to make the purchase. It felt as though we were getting so near but just couldn't close that gap and our time was beginning to run out. With blind confidence, we continued to push things forward with the solicitor, driving the purchase as much as we could before a point of no return was reached.

During this whole time, one close friend in particular took an interest, asking lots of questions about our plans, encouraging us to take stock and really assess if we felt 100 per cent sure about this potential new life-changing venture. She was a true pillar of strength to us in those days, providing endless support until, one day, I had a phone conversation with her that would change our lives for ever.

Friend: How are you getting on with finding the money?

Me: Well, not so great. We feel as if we've exhausted every avenue. We're not sure we can make this happen.

Friend: Well, I've been having a think and I'd like to lend you the money. I have a few things I'd have to arrange but I would be able to make enough available to help you make up the shortfall.

Me: What? No, no, that's too much. We couldn't.

Friend: No, I insist. If you want it, I will lend it to you.

How do you react after a conversation like that? One where someone effectively says: 'Here, have your dream. If you want it.' I didn't know how and neither did Sandra when I blurted all this out to her. While we wanted to scream YES, we were so shocked by this overwhelming offer of generosity that we just sat there, completely stunned.

This was a real, genuine solution. With every penny we had and now this incredible loan, we actually had enough money to make a realistic offer for Lynbreck. We spent the rest of that evening going over our figures, checking our maths and reviewing our planning, just to give us that final reassurance that this really, definitely, 100 per cent was going to work.

After a bit of to-ing and fro-ing, our offer was finally accepted and it was down to our legal team to wade through the complexities of crofting law. We started the déjà vu of resigning once again from our jobs and packing up everything we owned, something we were beginning to make into a bit of a habit. Just six months after our first visit, the deal was secured. We had bought ourselves an old second-hand four-wheel drive using some of the money from selling my car, anticipating the need for something more substantial than our two Polos. Then, for the last time, we filled the four-wheel drive and a rented Transit van full of our worldly goods before heading north. We arrived at Lynbreck just after 10am on 18 March 2016, located the key to our cabin under a nearby stone and unlocked the door to our new home.

CHAPTER 2

Crofting

A croft is an agricultural holding, usually small in size and averaging around five hectares, and it is unique to Scotland. Crofting has been a way of life here for many years as historically people worked small pockets of land with a mix of arable and pasture to provide food primarily for themselves, but living very much as part of a community. This traditional livelihood was the epitome of subsistence existence, where daily life was about survival in the truest sense and where the focus was on having enough food, water and heat to get by.

During the second half of the eighteenth century, a period of social unrest that became known as the Highland Clearances rippled through the north of Scotland. The rich landowners, who dominated the vast majority of the land, started to clear the tenant farmers from their holdings in favour of large-scale sheep pastoralism, which was deemed at the time to provide better financial return for less effort. The tenants were resettled to coastal areas onto small strips of land with poor soils, where they had just enough space to produce most of their subsistence from, but not quite enough to survive on. This often forced the people to make extra income from the fishing and kelp industries or from any work that was offered on estates, thus providing landlords with cheap, readily available labour.

The first half of the nineteenth century brought further hardship as the new settlements, known as crofting townships, were vastly overpopulated, resulting in widespread famine as they could not support the numbers of people that had been forcibly resettled. The problem was further exacerbated by a devastating potato blight, which started around 1846 and lasted for ten years.

The landowners began a programme of 'assisted emigration' in an attempt to depopulate areas by providing tenants passage abroad, particularly to North America and Australia. While this had fairly significant take-up, eventually the people on the land decided to fight back, signalling the start of the Crofter's War in 1882, which saw a series of violent conflicts between crofters, who wanted more rights, and the authorities and landowners.

To address the growing unrest, the British government appointed Lord Napier to undertake a review of the situation, which led to the Crofters Holdings (Scotland) Act of 1886. Lord Napier was widely regarded as having sympathy for the cause of the people and, through the new legislation, ensured that those in the legally defined crofting counties, located in the north and west of Scotland, would have the security of tenure. This meant that they could no longer be evicted from their land if they paid their rents and kept the land in good working order.

However, many crofters were unhappy that the act didn't give them all the rights they believed they were entitled to and crofting law has been subject to a number of changes over the years. Perhaps the most significant was the Crofting Reform (Scotland) Act 1976, where tenant crofters were given the opportunity to buy their land, the price dictated by the total value of the rent for a fifteen-year period. This gave a number of crofters even further security as they could become owner-occupiers of their land, an opportunity that had been hard won by them through decades of fighting for the rights they duly deserved.

Today, crofting law is mired in complexity and, while it gives protection to those that live on and work the land, in most cases banks will not provide loans or mortgages to assist with the purchase, believing that if the borrower falls into financial difficulty and defaults on payments, the Scottish Government will have first rights to the holding. This has resulted in house plots having to be de-crofted to sit outside of the legislation in order to attract the financial assistance that many require.

And for those that do manage to secure a tenancy or can become owner-occupiers, there are two main laws that must be adhered to. Firstly, they must live on or within thirty-two kilometres of the croft and, secondly, they must put the croft to use, a statement that has been widened to include a range of land diversification opportunities. If these two laws are not complied with, the Crofting Commission – a non-departmental government body that is the current incarnation of the commission that was established as a result of the 1886 act – has the power to take action to ensure that croft land does not suffer neglect of use or that the properties become absorbed into the blossoming second-home market. Sadly, the latter is becoming the reality as the commission rarely takes action, itself drowning in a mire of government bureaucracy; while crofts are increasingly bought up by those who have no real desire to work the land, instead drawn to the crofting counties by the breathtaking landscapes and a place to escape the ever-expanding urban sprawls to the south.

One of the biggest challenges faced today in crofting is making it profitable. Early on, we came across a lady called Joyce Campbell, a crofter based on the north coast who was very active on social media, sharing dramatic pictures of her land in Armadale, just thirty miles west of Thurso, and the stories of her day-to-day work. Joyce's family had been at Armadale Farm for many generations and, when it was her grandparents and father's turn to take over the land in 1962, they insisted on it having crofting status to access the security that the tenure would provide for the family.

Joyce recalls very fondly her childhood as a bairn (a Scottish term for a child), growing up in a vibrant rural community, full of big characters, where communality was part of everyday life, as people regularly came together to help with seasonal activities including sheep gathering and clipping, hay making and harvesting. While life had many challenges, it was the people and community working together and supporting one another that helped to get through the bad times and celebrate the good.

She explains that since she took over the farm at just twenty years old, life has very much changed. A fact no more poignantly illustrated than by her statement that, 'The school bus no longer picks up children from Armadale as it used to back in the day,' – a time when the population was flourishing. Like many crofting townships, the land is starting to fall into neglect as the population ages and the new, incoming blood to the area is heavily dominated by those snapping up the land as it becomes available on the open market for holiday homes, outbidding the few young locals that are left.

She shares honestly that it's not just the new arrivals who can pose a threat to the crofting culture, it's also the very people who have known nothing else that are sometimes struggling to accept and embrace new ways of doing things in a rapidly changing world. They are known affectionately as 'the old guards' – those who croft the way they always have done, or as they say, it's 'aye been' – a mentality that is embedded amongst many in the wider farming community as well.

But for new entrants and old guards to survive, they have to carve out a multilayered business model that takes the essence of crofting and makes it relevant for the twenty-first century, and there is no one doing this better and more effectively than Joyce. In addition to her sheep and cattle enterprises, she has a small flock of laying hens, whose eggs she sells locally, and she has embraced the tourism that has flooded the area by renovating the old croft house into holiday accommodation. Her savvy business mind, natural ability for marketing, and genuine warmth for people and crofting have elevated her profile to where she has attracted sponsorship from a range of businesses that financially benefits her day-to-day farming.

The approach that Joyce has taken is a perfect model that balances the injection of a fresh way of working to how she runs her croft but still retains the people element of the old guard 'aye been' mentality, as she plays an active role in fighting to keep her community alive. She talks of campaigning for investment from a nearby wind farm for a new community hall, standing behind a banner that says: 'We are the local people. We are the endangered species.' An additional threat to these communities is the growing call to clear the surrounding hills and glens of what some describe as ecologically damaging sheep farming, a problem usually caused by having too many animals on the land and poor grazing management. They want to return the land back to nature but have no real acceptance or understanding of the people who live and work there and the cultural pain that is still felt from the Highland Clearances of just a couple of centuries or so before. As Joyce says, we need lights on in the hills and glens, not the darkness of a landscape that people and their livelihoods have been cleared from once again.

But she does accept that to ensure its survival as an industry, everyone needs to be doing more for nature, criticising the consequences of the old European Union Common Agricultural Policy where, in the early days, farmers and crofters were rewarded financially for how many animals they kept under a system called 'headage'. The objective was to increase food production for a rapidly expanding population but it led to unforeseen damage to the land through overstocking, the point at which the land was carrying more animals than it was capable of supporting, and left a legacy of substantial environmental damage. Now that the UK has left the EU, Joyce calls for a change to agricultural policy that would reward crofters and farmers for doing more for the natural environment, a system that, as she puts it, 'literally pushes a different way of doing things down their throats' without compromising the ability to produce food.

Through conversations with people like Joyce, we were beginning to understand the complexity of economic, social, environmental and political challenges that had been faced by many in the crofting community both in the past and present day, becoming acutely aware of the impact of their struggles as many have now left the land. We wondered what our role would be in our new community, already having a huge respect for those that had gone before. What would we become?

CHAPTER 3

We Want to Work the Land

Lynbreck first appears on the Roy Military Survey of Scotland, known as the 'Great Map', which is a map of military roads drawn up by William Roy and dating to the mid 1700s. The road that transects the croft is known as the Old Military Road, running from Blairgowrie in the south to Fort George in the north and is likely the original site of an old drove road, an ancient route that would have been used by people to traverse the landscape and to move livestock. On the old map Lynbreck was spelled Lynbreack and was marked by five red dots which indicated habitation, although it is unsure if each dot represented an actual building.

In his book *The Strathspey Trilogy, Place Names Around Aviemore*, local historian John Halliday documents further changes in the name that include Linbrock in 1770, Lyne Breack in 1811, Lymbreck in 1841, Linbroch sometime between 1841 and 1891, and Lynbhreach in 1891 before eventually becoming Lynbreck or Lynebreck as it is spelled on the current Ordnance Survey maps. The name comes from the old Gaelic, whereby 'breck' means speckled and 'lyn(e)' means field or enclosure. John explains that when the first Ordnance Survey maps of

Scotland were being created, the cartographers would ask the local people for the names of places and write them down phonetically, often misspelling the native Gaelic into an anglicised version, a possible explanation for the variations over the years. But many of the locals pronounce it closer as to how it would have been many years back: phonetically Lyn – pronounced 'line' – and Breck – pronounced 'vhrecht'.

On the day of our big arrival, the cloud sat heavily on the Cairngorms, covering the mountain tops right down to the valley floor. Scotland is renowned for its charismatic landscapes. There are the jagged mountains of the Cuillin on the Isle of Skye, the watery expanse of Loch Maree dotted with islands that are home to ancient Caledonian forest and the beaches of the Outer Hebrides where the pure white sands slip into the deep blue water of the mighty Atlantic Ocean. We would now add the view from Lynbreck to rival any of those dramatic landscapes, where the flat open mire disappears into the deep green ancient pine woodland before culminating on the arctic plateau of the Am Monadh Ruadh (the Cairngorms), which, roughly translated from Gaelic, means 'the red mountains'. When the last glaciers retreated, the bare and exposed granite of the Cairngorm range would have had a reddish hue, in contrast to present day where the exposed rocks have greyed like tired ageing men resting in their deep earthy bed, and are covered in a patchwork sheet of lichens and mosses. But in March the view that had made such an impression on us was hidden in murky, misty depths, the previous owner having already warned us that everything looks a bit tired and washed out after a long winter slumber, with the landscape dominated by browns and yellows, making it feel quite raw and bleak.

The cabin was bare and empty but the wooden walls and worn red carpet made it feel warm, welcoming and homely. It had always been a dream of ours to live in a wooden house and stepping inside felt like walking into the heart of a tree. It had a unique, sweet smell reminiscent of seasoned pine needles roasting on a hot summer's day, a smell we inhaled deep into our lungs as we stood in the living room looking out across the vast expanse and into the mist. Once we managed to release ourselves from the hypnotic effect of the landscape in front of us, we investigated the empty cabin, its three bedrooms hidden in a wooden Tardis.

The two old buildings in the heart of the homestead stood as impressive relics and a reminder of a bygone people who once lived and worked here every day. The croft had been owned by the Grant family for over three hundred years, passed down from one generation to the next until the mid 1990s when it was finally sold. The roof of the byre – an old stone cowshed – was in a poor state of repairs with only half of the old tin sheeting attached, the northern corner of which would noisily flap in the wind. The deep orange hue of the rusted metal gave it a rather tattered look, casting a slight air of sadness over this neglected masonry masterpiece.

The old croft house was in a slightly better condition, with the roof and old grey slates keeping the worst of the driving wet weather from creeping into what used to be the original crofting family home. Inside it had been all but gutted, ready for a renovation project by the previous owners that never transpired, leaving it as an empty shell full of invisible ghosts from days gone by. We yearned for the walls to talk as we started to familiarise ourselves with every nook and cranny, the house full of features that showed the incredible workmanship and skill of previous craftsmen. On the southern end, we found an inscription above the old hearth that read 'J.G. 1852', later discovering that J.G. was likely John Grant, the date probably indicating the year the house was built.

In a darkened corner, we found an old table and chairs. After a bit of a dust, we carried them across to the cabin, gave them a clean and applied some wood glue to the legs and back supports of the chairs, which insisted on detaching themselves every time we tried to lift one. While to some people these would be destined straight for the skip or into a bonfire, they would do us just fine as our dining furniture and we were delighted with our discovery and first chance to upcycle.

It only took us an hour or so to unpack the van and four-by-four, after which we had a quick cup of tea and bite to eat before venturing out to explore the croft holding properly for the first time, realising how little we actually knew about the land we had just spent our life savings on. We decided to head for the highest point, the peak of Beinn an Fhudair, which translates from Gaelic as 'hill of the powder', the summit of which we shared ownership with two neighbouring landholdings. This, we thought, would give us a bird's eye view and reference point of where Lynbreck sat in the surrounding landscape.

At the summit, our attention was drawn to the many little houses dotted across the landscape, some isolated, others part of a little group and some as ghostly empty shells with just a stone skeleton left behind. Today, these mostly abandoned settlements give the impression of desolate isolated dwellings but in their time were busy micro-centres of human life and activity that were connected by a strong and vibrant social and community network. John Halliday comes from a long line of crofters and pinpoints the 1960s and 1970s as the time when things really started to change culturally; when the young began to move away in search of a more prosperous and easier life, and mechanised equipment took the jobs of many men. He remembers harvest time, a real communal gathering when oats, a commonly grown crop of the area, were harvested collectively, the sheaves gathered into stooks and then taken into the barn for threshing. The whole day would be spent out in the fields, working together, stopping only to have their 'piece', a Scottish phrase for a packed lunch, which would have been prepared by the crofter's wife. Another part of the land often had a late crop of turnips, known locally as neeps, which would have been used as an important source of feed for overwintering livestock. This also required a large communal effort where armies of extended families would be out preparing, planting, tending and then harvesting the crop before the onset of the harsh cold.

The winters in those days were dramatically different to the conditions we experience now. The records from the old school house on the Dava, a large expanse of moor and hills to the north of Grantown-on-Spey, document that snow could fall any time from October to May. Some days it was so bad that many of the children who walked there from the surrounding areas simply could not make it through and it was not unknown for isolated farms to be inaccessible by road for up to six weeks. In December 1880, a trapped train on the Dava line was buried by 77 feet of snow when it was eventually found, and in February 1963, drifts of 30 feet on the same line took 49 men with dynamite over 21 days to clear.

John describes their way of living as very seasonal, and explains how their tasks and commitments would change in synergy with the land, climate and community rhythms around them, as the year transitioned from cold, dark nights of hibernation to warm, bright days of activity. The highlight of summer was the Grantown Show, an event that attracted the whole community to join together and celebrate the local agricultural traditions.

Farmers and crofters from the surrounding areas would walk their best livestock to the showground, occasionally losing a few on their way there or way back, in the hope of winning one of the coveted rosettes for their finest specimens and maybe making a few sales or purchases. The women would bring baking and craftwork, entering categories for the best scones, jam and honey as children lined up to take part in the travelling fairs that would arrive to provide the entertainment.

After the show, the next big date was 12 August, known as the Glorious Twelfth, and indicated the start of the grouse shooting season, where the wealthy landowners would turn out in droves to shoot the native red grouse, strongly supported by the local community. John talks of how the young folk would be employed as beaters, a team that would walk in front of the shooters to flush out the grouse from the heather into the sky, getting paid £5 a day and a can of beer. The main agricultural harvest would precede the pheasant shooting season in autumn when thousands of birds were released into the countryside and then shot, providing a meat not too dissimilar from chicken. And, shortly after these outdoor events, it would be time for Christmas and New Year celebrations, the latter known in Scotland as Hogmanay and associated with some strong local customs. It was common practice to visit neighbours to 'first foot', a tradition whereby individuals or groups would arrive at the door, receiving a dram of whisky and a bite to eat to celebrate the dawn of a new year before moving on to the next house. In rural communities, it was normal for people to walk, cycle or even drive miles between residences, as first footing could go on for a few days. And, as the bells struck midnight, the farmers would go out with their guns and fire a shot up into the sky, a precursor to the fireworks of modern-day celebrations, which would have had no less of an impressive effect as the silence of the dark night sky would be blasted by the chorus of gunshots and the scent of spent gunpowder would fill the surrounding air. It must have been a very comforting sound, almost as if each family was shouting 'Happy Hogmanay' with every shot, dispelling any feelings of loneliness or isolation.

On New Year's Day, the community would get together for the annual hare hunt, heading into the hills to shoot the iconic white hare, always bringing home their kill to go into the pot and never taking more than they would need to feed their family and friends. And, after all the celebrations, it was time to hunker down for the long stretch of winter that would eventually turn into spring, when the whole seasonal dance would start again.

Today, life in the area is somewhat different as most small farms and crofts are no longer farmed with the same community involvement, and many of the old traditions only exist in fond memories and stories shared of days gone by. And, as our focus shifted to the community that lay around us, we couldn't help but wonder how we would be welcomed as new faces to the area. This is a part of Scotland that is popular with holidaymakers and second-home owners, where a mixture of wealthy individuals and organisations own vast swathes of land. The result is a place where the local population is being squeezed to the edge, no longer able to afford the inflated housing and land prices in an increasingly popular area made even more attractive by the relatively recent designation of National Park status.

National Parks in the UK are very different to many similar designated areas around the world in that the National Park authorities – the independent government-funded bodies responsible for looking after each park – do not actually own any of the land within the park boundary. The role, therefore, of the employees is to heavily engage with the owners to ensure the aims of the National Parks are being delivered, a task often easier said than done when landowners are as diverse as farmers and crofters, conservation and rewilding organisations, as well as privately owned estates managed for grouse shooting, deer stalking and commercial forestry. This complexity of ownership can inevitably result in clashes as different economic, social and environmental agendas collide, leading to a disjointed landscape of visible and invisible barriers where the fluidity of nature and her systems become interrupted by fences, walls and mindsets.

We were acutely aware that this title of landowners could now be applied to us and we had to reflect honestly on our decision and ability to buy the croft and whether we had deprived a born-and-bred local of the opportunity to take on the land. While the croft had been on the market for over a year, it's a question that will always go unanswered with all the possible unknown consequences, the shadows of which we will always have to live with. While technically we were incomers, a term used to describe new people moving to the area, we hoped our contribution would be to share our own ideas and fresh energy to enrich the nature and community around us.

We met our three nearest neighbours first, all farmers and crofters with longstanding family connections to this area and a real bunch of hardy folks who had spent years working the land, trying to carve out a decent living to support their families in one of the toughest, harshest and most unforgiving environments. They were full of questions in those early conversations, which in general went something like this:

Question: Where are you from?

Me: Oh, well, I'm from Northern Ireland, and Sandra grew up in Switzerland but her mum is Scottish. [We liked to get this in early on to show there was at least some family connection to Scotland.]

Question: Where have you moved from?

Me: So, we were down in the south of Scotland, and before that we lived on the edge of London and worked as rangers for the National Trust. [It's a bit of a cliché that people from the affluent south-east of England escape to the Highlands and we really didn't want to be negatively judged in that way.]

Question: So, what are you going to do with the place?

Me: We want to work the land.

The last answer was always well received. The people we met were kind and genuine to the core with a deep-rooted connection to their land, their animals and those around them. But they were also smart, savvy, hardworking business owners who knew how to thrive in adversity while always appreciating that it's the simple things in life that mean the most.

We came away from those early meetings surrounded in the warmth of their welcoming glow and feeling an overwhelming sense of relief for the open arms and hearts with which we were received. And, from day one, we never hid the fact that we were a couple. The opinions on same-sex relationships have certainly moved on from where they were even a few decades ago, and often the impression is that areas such as these may hold more prejudiced or old-fashioned views. But never once did we face any discrimination or made to feel any different to anyone else. From that point on, we became known as 'the girls' as this new and unknown community started to become 'our' community, and we were determined not to let them down.

The memories of those first few weeks in our new roles as custodians of Lynbreck Croft are as fresh and clear as the view of the Cairngorm mountains on a bright sunny day. It was a time we had dreamed of for many years, wondering how it would feel and what we would be doing when this day came. The reality was unexpectedly different. We dreamed of living on the land, working the land, but what did this actually mean? Finding the croft had been beyond a dream come true, where the goal of a five-acre plot had grown fifteen times to the one hundred and fifty acres that made up the entire Lynbreck parcel. When we sat on those long, dark winter nights during our time down in the south-east of England, brainstorming words onto pages of what our new life might be like, it was always on a much smaller piece of land, not something of this size. And, even when we found Lynbreck, our thoughts had been so consumed with pulling together the money to buy it that any early future plans focused on finding external work to bring in the income we needed in the beginning to pay the bills. We had spreadsheets with figures that showed how much we would need to earn to pay the council tax, home and car insurance, phone line and internet, fuel and food, and the easiest way to acquire that money would be from a job.

It's quite incredible that even at this stage, full-time farming was not really on our radar – even though the size of our landholding had grown, our initial vision of growing our own food, keeping a few hens and maybe the odd camping spot for a bit of income, stayed much the same. It's clear to us now that in spite of all the early planning, thinking, researching and endless spreadsheets, the reality was we didn't actually know what we were doing other than following our hearts and our dreams.

CHAPTER 4

In the Beginning

We would awaken each morning with a start, the realisation of where we were jolting us up and out of bed, then swiftly inhaling breakfast in anticipation of the day that lay ahead. The problem was that we didn't really know what to do, and so those early days were often filled with wandering around, looking for a new sense of purpose and feeling a little lost in our own home as we tried to fill the void of time between getting up and going to bed.

In normal life, people have a routine whereby they get up, go to work, come home, eat dinner, sleep and repeat, followed by a weekend of relaxation and down time. This routine provides structure and consistency and, while sometimes can feel tiring and monotonous, it also creates feelings of purpose and security. Our new situation was the opposite, lacking everything a routine provides, and we had to adapt quickly to a life that was completely different from anything we had experienced before.

But as we familiarised ourselves with the land, we realised just how much work there was to do. Everything from ripping out miles of ancient fences where there was more wire embedded in the ground beneath than between the rotten posts, to gutting and renovating the old buildings – neither of us having any building experience or skills but desperately wanting to make some kind of a start. In fact, there were so many jobs that we had no idea which to tackle first. What was the most important? Where would we even start? I'm a big fan of sitting down with a cup of tea to take stock of life and draw up plans but even I was starting to get itchy for some action and wanting to skip the tea drinking – a fact that coming from Northern Ireland where tea is a staple drink and the resolver of all problems, was a bit of a worry.

In between sitting still and wandering aimlessly, we had noticed a number of trees in and around the homestead that needed a little attention, so we finally made a decision, as a first job, to spend some time tending to the young saplings, weeding out competing grasses and replacing broken tree guards. Psychologically, it was the perfect task as we felt comfortable and comforted in the familiarity of working with trees, and it was the kind of job that you can step back from and see what you've done. It was little wins like these in those early days that kept us ticking along, acting as a confidence-boosting springboard to the next one.

After sorting the trees, we decided to tackle where to build the new kitchen garden, which was part of our original vision for self-sufficiency. The croft maintenance jobs that we had started to make a list of could happen alongside, but this became our priority, wanting to get things ready for the approaching growing season. Our plan was for a 'no-dig' vegetable plot designed following the basic principles of permaculture, a way of efficiently growing in harmony with natural systems. The no-dig approach advocates keeping the natural soil structure intact, avoiding turning the soil over, which can lead to greater weed growth and damage to the biology below. Permaculture design incorporates methods that seek to make growing both easy and abundant by choosing the right site, avoiding the use of all chemicals and prioritising soil health. Permaculture gardens are often stacked with many levels of food crops, where plants that have beneficial relationships are kept together and other aspects of nature are encouraged, such as borders of wildflowers to attract pollinators and predatory insects who prey on other life that may want to dine on our establishing mini food forest. While we'd done plenty of reading on permaculture and no-dig and it all sounded great, neither of us had actually ever designed a kitchen garden from scratch ...

Directly in front of our cabin lay a south-facing, flat, grassy plot that would provide an ideal location. Within just a few days we had erected a very simple rabbit-proof fence, and marked out and built five raised beds. These we filled with loose soil from the tops of mole hills, an arduous, backbreaking task that involved trundling up and down the fields with a shovel and wheelbarrow, but a job totally worth it for the crumbly, rich topsoil. A neighbouring farmer friend took pity on us, arriving one day with a tractor bucket-load of cow manure to mix in with the soil and build fertility, the kind of warm-hearted gesture that became the signature of that shown to us by our surrounding community.

Another early job involved spending a day in our woods, felling some trees for firewood as we also wanted to be self-sufficient in fuel for heating. With the last of our money, we had invested in a small multifuel stove, which would become our only source of heating in the cabin. Sandra's dad had bought us a new chainsaw as a house-moving gift and we were so looking forward to getting it out and using it. Although we'd worked with trees for many years, it had always been using an employer's equipment in our employer's woods. This time, however, it was our equipment in our woods, and that knowledge alone gave us such a thrill. But, in spite of our enthusiasm, we didn't want to go too overboard, selecting and felling a modest number of birch trees, trimming off the side branches and chopping them into lengths that we could comfortably carry on a number of trips over our shoulders the few hundred metres back to the homestead.

In between spells of action and adding to our growing list of tasks, we would sometimes find ourselves sitting outside, acknowledging the silence of the surrounding countryside. The croft is transected by what can be a busy tourist road but, during the evenings in particular, the traffic is minimal. That depth of silence can be quite a strange sensation to get used to and one that is a contradiction when we say that the peace and quiet around us was also really loud, broken on occasion by the sound of the wind in the trees or the chatter of a foraging robin. As our ears slowly adapted to this new normality, we began to experience what it feels like to just 'be' rather than always 'do'. With these moments of just being, we began to feel a sense of safety and comfort in the embrace of nature that was all around us, and each task completed began to feel like a scaffold pole. As we started to finish one and begin another, we were connecting the poles in a way that formed the beginning of a framework from which we could build our new lives.

After our first three weeks at Lynbreck, progress was definitely being made as we settled into a new rhythm of life on the croft. I had found a four days a week job based fifty minutes away in Inverness that would give us an instant income to pay the bills, so it was soon time for Sandra to take a lead on day-to-day operations while also looking for part-time work locally to cover any additional shortfalls. Eventually she managed to get a contract on a local estate, strimming back the vegetation on the banks of fishing beats along the River Spey. The job started at

5am and finished at 9am, allowing the area to be clear and well presented for the day's fisher men and women to come and enjoy the tranquillity and beauty of the riverside. That summer, she picked up more work strimming sections of the Speyside Way, a long-distance walking route running from the Moray coast to the edge of the Cairngorm mountain range. It was a job that involved miles of walking and carrying heavy equipment, a can of fuel and her lunch in a backpack. While these jobs 'worked' at the time to provide some income and were based in stunning locations, they were anything but glamorous, the success of the day being measured by how few hidden dog poos came into contact with the machine – and subsequently poor Sandra.

It's a tricky balance that we now see most new entrants to farming face. On the one hand, it's essential to keep the money coming in but, on the other hand, you need to be dedicating time to planning and building your new business. The latter is something that takes a huge amount of time, focus and energy, and it was a frustrating catch-22 situation when we were first starting out, which we couldn't find a solution to. And the reality was, and is for many today, that there simply is no solution. Unless you have a separate pot of savings to draw on, the financial pressures are too great to simply go it alone from day one. We had monthly bills like council tax, electricity, car insurance – things that we simply couldn't delay or defer and that we needed an immediate income to cover.

On the days when we were both at Lynbreck, our individual skills and talents began to shine through, naturally falling into different roles or tackling particular jobs. Sandra excelled in all the outdoor tasks, nurturing innate abilities in joinery and carpentry carried through her family genes, coming as she did from a line of natural craftspeople. Sandra has the mind of an engineer, which shines through in meticulous planning and design, complemented by Swiss characteristics that surface in her attention to detail and high standards of work, building the small-scale infrastructure that was essential to meet our basic requirements as work progressed. Meanwhile, my talents lie with words, writing and desk-based tasks as well as talking, a characteristic I ascribe to my Northern Irish heritage and culture where the ability to chat to anyone about anything is a national trait; one that played an important role early on as we looked for help and asked for advice. But I would regularly rebel if stuck inside for too long, craving jobs outdoors to get some fresh air and exercise. I was, quite randomly, rather a dab hand with the chainsaw and very useful at any jobs that required the demolition of old infrastructure, returning from ripping out endless fences with a real sense of satisfaction.

Yet while we started to fall into some sort of a routine with our croft work, our emotional state was anything but constant. When the sun shone, life was good. When a little job was ticked off the list, life was good. Our fierce determination was partnered with a kind of manic energy. When we finished those little tasks like weeding trees, we would think and say, 'Yes. We are doing this,' followed by a high five, kicking any concerns or worries aside as we were taking on the world and winning. But when another unexpectedly hefty bill would come in or we failed spectacularly at a job tackled, we were crushed, feeling despondent and downhearted as our confidence dived.

'It tastes really strange, almost like garlic,' I said, nearly gagging from the sip of water that had come directly out of the tap. 'I think we might have poisoned our own water supply.'

The previous owner had left instructions on how to clean and sterilise the well, something that seemed like a perfectly appropriate and sensible thing to do. Sandra took a trip into Inverness and, with a bit of research and advice from a sales person, found a cleaning agent that would do the job, being very careful to select something that was not so nasty and horrible that it would cause a mini environmental incident. We bought an extraction pump and, following the instructions carefully, drew all the water out from the well, filled the empty space with the cleaning agent and then allowed it to refill. The plan was that we would leave it a while, pump it out, repeat that a few times and, voila, we would have a nice clean well.

It didn't happen like that.

After a few empty-refill-empty-refills, we ran the water through the taps in the house but something was just not right. We decided to leave it a few days, during which time we switched back to drinking water from containers that I would refill from the houses of friends and at my work, but things didn't improve as the well water had the strangest metallic smell and taste. After a few weeks I called a local water specialist, explaining what we had done. I have never felt so stupid in all my life as I tried to explain how we had tried to clean out what we now realise was already a beautifully naturally sterile well with a load of chemicals. The only point at which we had veered from the original instructions was that we didn't use bleach, something we wanted to avoid for its detrimental environmental credentials, so there must have been something in that cleaning agent that reacted badly. The long and short of it was that we had been trying so hard to do everything right, and do it independently to prove to ourselves that we could, that we ended up poisoning our own source of water. A stunning fail. Luckily, within a few months it sorted itself out and the regular marathon of refilling our drinking water bottles came to an end and we vowed never to consider doing it again.

This was just one of the many mistakes we made in those early days as we tried to find our feet. And if it wasn't for the little unexpected boosts and generosity from new, friendly neighbours, it might have all gotten too much. One day, while working in the kitchen garden, the silence of the croft was punctuated with a growling rumble that grew louder and louder. Our heads peeked above the grassy bank as our eyes took sight of a man coming down our track, rolling what looked to be a large wooden barrel in front of him.

'Hi, I'm Donald. I'm your neighbour,' he said before presenting the old whisky cask he had brought as a gift from our local distillery in Tomintoul, sadly empty of its original contents but a beautiful example of the craftsmanship of a skilled cooper. As we chatted with Donald, it soon transpired that he actually lived ten minutes away by car, so not a neighbour in the literal meaning of the word, but that didn't matter as we learned that the exact geographical distances between those of us in rural settings are often inconsequential. He was born and bred of this area and his warm, kindly manner, big smile and wicked sense of humour made us feel beyond welcomed. From the time he arrived to the time that he left we had transitioned from new acquaintances to firm friends.

Another neighbour Fraser announced that he would soon be bringing us up a bucket of peats for the multifuel stove we'd just had installed. A traditional crofter, he still cut his own peats for the fire from a boggy area, which he and his brother would stack by hand into cone-shaped structures to dry in the warm summer sun. We imagined a hand bucketful, but he meant a tractor bucket, which we realised when he appeared one day, tipped the peat onto the grass next to the old byre without any fuss and went away about his business. The next year, he brought us three tractor bucket loads, the growing volume reflecting the growing affection and blossoming relationship that was starting to build with this new network.

The extraction of peat is very much frowned upon today as the intensified mining and harvesting have decimated landscapes and precious habitats, emitting millennia worth of stored carbon. But the way in which they would do it, only taking what was needed for themselves like their many forefathers had before them, harked back to a time when humans had much greater respect for the land they lived and worked on. They shouldn't be tarred with the same brush as the big polluters at the other end of the spectrum, where land is purely exploited for financial gain, irrespective of the consequences.

These are just a few memories of many experiences that impressed on us early the importance and value of community in the rural landscape. In our area, people from the land don't phone, text or email to make an appointment for a visit, they just call round for a chat and catch up as they are passing. There is an unspoken rule where if you have time, you offer a cup of tea but if you don't then both sides know, intuitively understanding that it's a busy period or just doesn't suit. It's never a strange or awkward situation and no one ever gets offended. On some days, no tasks were completed as neighbour after neighbour would call in to say hello, offering very genuine help and support should it be needed. For Sandra this was a completely new experience, not at all like the more reserved modern-day Switzerland but, for me, it felt entirely normal and reminded me of the Northern Irish culture where I came from. In contrast, there would be weeks and months go by when everyone was busy and no one called, including us, but that also didn't matter.

When we talk in society about the importance of protecting rural communities, it is this that we are protecting. A way of living that brings out the best in people, who steady you before you fall or pick you up when the ground disappears from beneath you. Where the strands of loyalty and kindness connect us in an invisible web that weaves across hills, over rivers and through forests. I spent over ten years of my life living in busy, bustling cities where I didn't know my neighbours and often felt lonely, in spite of being surrounded at close quarters by people 24/7 so, for us, it is a very special feeling to look out across the landscape on a dark winter night and see the lights of the scattered houses, twinkling like flares of friendship through the dark, acting as a constant reminder that should help be needed, every one of them would be there to lend a hand. We were no longer a lone partnership, instead players in a team, committing to pull our weight and contribute positively to our collective role in keeping this community alive. And as we started to bed in to our own landscape and roles on the croft, we began to imagine what our new business at Lynbreck might look like, accepting it would be a dramatically scaled-up version of the idealised vision we had dreamed of many years before.

CHAPTER 5

The Call of the Ancestors

It wasn't just neighbours who were calling by with increasing frequency, but members of the original family who used to live at Lynbreck, many of whom had not been back for twenty or thirty years. We would wander with them around the homestead and into the old buildings as they told the stories we had so longed to hear. In the now disused croft house, the downstairs had a galley kitchen big enough for just one person, with a tiny window that looked due west towards Abernethy Forest, a little cupboard-sized room with a toilet so small that it required the user to reverse into it, and a living room with a large range that would have pumped out heat day and night, combatting the constant chill that often comes with living in a stone building without any insulation. Opposite the range and behind the sofa in the centre of the house sat an old box bed that was used by the family when there were up to eleven people living in this tiny residence.

It was said that under this box bed, a feature in itself that was referred to by almost all of the old family that visited, below the floor was the entrance to a tunnel that ran beneath the building down into a steep-sided woodland gully about one hundred and fifty metres to the west of the house. On the 16 April 1746, the Battle of Culloden took place approximately forty miles to the north-west of Lynbreck, where the Jacobite army was brutally defeated by the British government forces who outnumbered and outmanoeuvred the exhausted and nearly starving fighters, bringing the Highland and Island clan system to an end.

One of the Jacobite leaders, a local man called John Roy Stewart, an acquaintance of the famous Bonnie Prince Charlie, survived and managed to flee the scene, eluding capture as he headed for his native Strathspey. One of the Lynbreck legends states that during his escape he took cover in a cave located in the steep ravine that runs from Grantown-on-Spey and ends at Lynbreck before spilling out onto the flats at the base of the Cairngorms. Before continuing on his travels, the story goes that he was fed and watered by the crofting family as they smuggled the goods from the house via the tunnel to the secret hiding place.

At the other end of the house was the good room, another small living space that was rarely used except for special occasions and visitors or to get food that was stored in the large wooden cabinet that ran across one wall. A small flight of near vertical stairs led to a narrow landing with a bedroom at either end. One of the last inhabitants, Gordie Grant who lived at Lynbreck until he was nineteen, told stories of just how cold it could get up there during the winter time. How in his parents' bedroom, the frost would gather on the nails of the exposed roof above, where only lengths of wood covered in slate separated them from the outside and the little roof light windows would be sealed firmly shut with ice.

The Grant family were active crofters right up until the point they sold Lynbreck in the late 1990s. Gordie told us how his mother was very fond of her cows, keeping a small herd of ten Blue Grey cattle, a common cross-breed well suited to Lynbreck for its hardiness and ability to graze on native vegetation. She also kept hens in a small stone shed that had been expertly crafted into the hillside, as well as geese and ducks that would roam around the homestead, greeting some of the many visitors that would call by and giving the place a sense of vibrancy and life that is so often associated with working farmyards.

The old stone byre was used as a garage for tractors and equipment, to store wood and to house a generator, which powered the lighting until the house was connected to the mains electricity network in the late 1980s. At the other end were two wooden framed stalls, used for calving or for housing the cattle in some of the winter months. Even in the 1980s, heavy snowfall was still a reality as Gordie told us about the time he had to dig a corridor through a wall of snow the height of the porch roof just to be able to leave the house. Or the time the snow was so bad they couldn't leave the croft for nine days due to the drifts that had blocked the main road into town.

Other older stories talked of sheep on the croft as well as an old Clydesdale called Jock, a working horse who would pull the plough or help with extracting timber for processing into firewood for the range. Jock knew his way around the croft and he would be sent, unmanned, from the woods to the house, dragging his heavy load behind up a steep and unrelenting hill, only to be relieved of the timber at his destination before being sent back for the next. And typical to crofting, the man of the house would often have another job to keep the money coming in. Gordie's dad worked on the railways and his father before him on the roads, both hard labouring jobs that required a lot of physical effort for little financial return.

We started to build a picture of a family who carved out a life and a livelihood for themselves in this beautiful but unforgiving landscape; people who were resilient, hardy and resourceful, making the most of what little they had. Some of our favourite tales recalled the parties that would happen in the house where the whisky flowed like water as the sounds of high jinx and laughter spilled out with the warmth of the light through the tiny single-paned glass window into the dark emptiness outside. And then there was the opening of the Lynbreck motorway – the time when the main entrance track was straightened and resurfaced to make access easier. This was such a big event for the family that they even held an opening ceremony where a red ribbon was cut to mark the official opening.

On some level, we started to connect to the previous inhabitants, understanding their hardships, which could have taken a heavy physical and mental toll, but respecting their spirit, still able to smile in the face of relentless adversity and take merriment from the social and cultural connections with the people who surrounded them. As we grew to know them, we felt that their legacy was a part of our future, promising to keep their story alive out of respect for who they were and what they did, and using the lessons we had learned from what they had achieved and how they survived to feed directly into our evolving new model for living at Lynbreck Croft in the twenty-first century.

Our instinct was to work the land to provide food not just for ourselves but for our wider community, all the while ensuring natural diversity could thrive. A way of farming where every decision was taken in the context of working with the land and the environment. And, personally, we wanted to build and strengthen our own relationship with nature, drawn to a more hands-on way of working the land and one of the many reasons we fell in love with the old crofting culture. It was a way of living and working where farming was about producing food within the natural capacity of the land, in rhythm with the changing seasons, when communities worked as large family units, and where loyalty and kinship were thicker than blood alone.

We were conscious to avoid romanticising every element through rose-tinted glasses and we had no illusions that the old conditions were anything other than harsh. We had no desire to live the original crofters' struggle of centuries before and, in all likelihood, would not have been able to endure it. We liked the comforts of modern-day living; a warm house, an electric blanket on a cold night, a decent internet connection, a reliable car and, while we saw ourselves as hardier than some, we didn't want to give up all of the benefits that many of us have become used to having.

Yet we looked back to those days with a huge amount of admiration for the crofter spirit, people who were able to live and survive just by working with what they had in a location and on ground where everything was stacked against them. And it was hard not to feel instinctively drawn to the way the older generation talked about the past, when farming was a community effort, not a one- or two-man band, and where everyone worked with nature because, again, that is all they had. While their gratitude for modern-day efficiencies and appliances shone through, their sorrow for the loss of that kinship was also clear; a bright flame that once burned brightly inside, but that had been reduced to a dull flicker as machines and mechanisation took away from man power and team work.

Sometimes it was as if we could now hear the ancestors of the empty ruins calling us to keep the spirit of Lynbreck alive, something that in our very core we could not ignore. But with buildings derelict, fences defunct and other key infrastructure like water pipes in the fields missing, where would we even begin? And what about us? We hadn't come from farming families or been to agricultural school, so could we actually do this?

One option would have been to carry on with our original plans: of self-sufficiency, hens and some camping plots, and leave the rest of the land to go wild. Just because we had more acres didn't mean we had to do more with them. But there was something that felt inherently wrong in that. It wasn't just about the tradition, about the land that had been worked for centuries by those before us, and continuing their work. It was more than that. It felt as though this call to the land was becoming our life calling, that there was something bigger at work here where our role would be to produce food from the land in a way that was in harmony with the nature we shared this precious space with. We weren't here to impose our will, but to find our place, to be guided by nature, to observe natural processes and work with them.

From our previous experience in conservation, we knew that large herbivores like aurochs, wild ancestors of domesticated cattle, would once have roamed the land, playing their role in the web of life. If we had cattle, could we work with them to be like aurochs? And wild boar would once have been more widespread, the disturbance they would cause creating spaces for new plants to colonise and grow. If we had pigs, could we work with them to be like wild boar? And what about our role in all of this?

A picture started to emerge in our minds, one that became clearer day by day, of a place where people and nature could thrive as one. Where decisions were made for the benefit of the whole, for the long term. It was an image that excited us, that intoxicated our minds to the point where it was all we thought about. But how on earth could we make it happen?

Agriculture in the UK is a heavily subsidised industry, as it is in many parts of the world today. We decided to investigate what funding might be available for our situation, eventually coming across the Young Farmers Start-Up Grant Scheme, provided by the Scottish Government and designed to help new entrants like us into farming. It required filling in an application form with a five-year business plan and, if successful, came with a financial investment of seventy thousand euro, of which 90 per cent would be paid upfront. There was only a small window of a couple of months to apply, so that summer we began to immerse ourselves in all things farming, trying to soak up as much information from as many avenues as possible.

The first challenge was to master the language of farming by learning the many agricultural words and phrases that are commonly used, yet were completely new to us. In sheep speak there are tups, gimmers, ewes, hogs, lambs and wethers. In cow speak there are bulls, cows, heifers, stots/steers/stirks/bullocks and calves. We learned that a sheep was no longer just a sheep and a cow was no longer just a cow as we navigated our way into this new world of language and culture.

Sometimes we felt confident enough to admit our ignorance and enquire about the meaning of a word, one time sitting down with our neighbour Fraser around our kitchen table and asking him to explain them all. At others, and with people we didn't know, it was more awkward, not wanting to look silly and desperately trying to fit in as farmers around us would talk between themselves and we would just stand and nod, but not having an actual clue what they were talking about. Those were the kind of moments when the worry of being asked a question or for a comment made our cheeks flush, knowing that if it happened, we might be exposed as the imposters we felt we were.

I spent every spare moment sitting in front of the computer, pulling together a business plan that set out what we wanted to do and was presented in a way that was as professional looking and sounding as I could make it. It was a very competitive grant scheme and this was the first time that either of us had ever had a go at writing such a document, but now we were starting to see what our new plans looked like on paper – anything we had drafted before was loose and vague on details, based on a much smaller imagined landholding that did not involve farming. We hired a specialist to help us with the financial forecasting and were able to use standard industry figures to help us estimate our expenditure and income over the coming years. Our plan was to sell produce direct, rather than through markets or other sales channels to maximise our profit. It was clear that our margins would be incredibly tight, and if we had any chance at all of making enough money to pay the bills, we'd have to take on most of the work ourselves. Not only were we to become farmers, we needed to become accountants, sales women and marketing experts as well as builders, fencers and growers. We'd breed all our own cows, all our own pigs, have twenty beehives, an army of hens, become accredited for everything including organic, turn all of our produce into something amazing (didn't know what, but it would be amazing), build a new barn, renovate the old byre, apply for more grants if available (successfully) and run a totally profitable new business (with zero experience), and all within a mere five years. It was a 'sky's the limit' kind of document. But, while it had a lot of information, the detail as to exactly how it would happen was fairly limited. We didn't talk about how we would graze the cattle or how we would work the pigs, simply because at that stage we didn't know. It was a plan that detailed enterprises and finance, rather than techniques or systems, with long shopping lists of all the things we needed to buy like a quad bike, fencing materials and equipment to help with safe handling of animals.

By the end of the summer, our business plan was complete and submitted just in time for the October deadline. During the next six months we continued to work both off and on the croft, planning yet more projects, the central one involving the planting of a new woodland. We never actually believed we wouldn't get the Young Farmers Start-Up grant, even considering it was such a competitive scheme and our plans were not entirely mainstream. We didn't just plan to be beef farmers or sheep farmers or pig farmers. We were talking of a diverse, multi-enterprise, financially viable small-scale unit, what is in effect the cultural essence of crofting, which has always been about getting little bits of money from lots of different income streams. And I don't think anyone actually believed it could be done in a way that would bring in enough money for us both to live and work here full time. If farms that were ten times our size couldn't manage it, how could we? But while there would be a lot of groundwork to be done, our figures said we could do it.

And then one day in April 2017, just a year after we had moved to Lynbreck, the letter arrived saying the money was on its way.

Someone, somewhere, thought we could do this. They had looked at our plan, believed in our (admittedly ambitious) goals and thought, 'Yes, I'm going to give these women a chance.' And what an enormous boost of confidence it gave us. Now we had the investment to get started, with an officially approved plan marked by the stamp from the Scottish Government. But what we were about to learn was that while the best-laid plans, particularly in farming, are all well and good, ours were very definitely going to change.

CHAPTER 6

Haunted by the Ents

We had bought the croft from a local family who had been there for around twenty years, and I remembered an early conversation with the dad who said: 'We want to see someone take the place on and do something with it. We didn't do anything.' In some respects, this was accurate as they never made that leap into actively working the land, meaning we purchased what could be described as an agriculturally semi-derelict croft with no functional infrastructure. But in our eyes, what they did do was everything that Lynbreck needed at that time. They allowed it a period of rest and respite from centuries of targeted management by humans. They let the grasses and wildflowers seed, allowed trees to come back and let nature begin to recolonise and take the lead. We now realise and deeply value how much they did do before handing the reins to us.

The best way to describe Lynbreck in just a few words is 150 acres of pure Scottishness, situated on the leeward edge of the Cairngorm mountains with a mixture of grassland, woodland, heathery hill and bog. There are three main fields, which we call near field, far field and lower field; the near field being the one around the homestead, the far field sitting to the south-east and the lower field at the bottom of the slope to the south-west.

The fields are separated by a band of largely connected woodland, which we refer to as middle woodland, lower woodland and far woodland. Near and far field and all the woodlands sit on a slope with the exception of a small band of flat land no more than twenty metres wide and a few hundred metres long that snakes around the contour in near field and includes the homestead. The heathery hill ground takes up the north-eastern and north-western corners with the bog completing the landholding in the southern section.

Our first exploration of the nature of Lynbreck was on that hazy summer day in August when we first came to visit. We needed to ground ourselves and gather our thoughts as well as better acquaint ourselves with this new land, so we took a walk across the field, the one we now refer to as near field, below the house.

It was full of long grass with giant seed heads and big, hidden tussocky bases requiring careful navigation. As we headed downhill towards the woodland, we soon found ourselves in a jungle of head-height bracken, paying little attention to what species were around us. The variety of pasture plants beneath our feet, as we later came to realise, were fairly representative of all our fields, known in farming speak as our in-bye fields, which relate to those closest to the homestead.

When we finally bought Lynbreck just seven months later, we wanted to take the time to really get to know the habitats and species present. It was imperative that our new agricultural operation would help to increase biodiversity across the croft and be driven by a deep respect for the natural world where our actions would contribute to more of all life and regenerate our land and our soils, rather than be extractive or depletive. And the only way to monitor it accurately would be to compare the land after a few years of farmed impact to what had been present before.

During our time in the Scottish Borders, we had become acquainted with a lady called Diana who worked as a professional ecologist and lived nearby. We hired her to undertake a baseline vegetation survey and, during just two days, Diana was able to identify 148 species of trees, shrubs, grasses, sedges, rushes, wildflowers, ferns and mosses, as well as providing some recommendations as to how to maintain and improve diversity in specific areas.

A large part of the Cairngorms National Park is made up of a jigsaw of nationally important habitats, which include dry heath, Caledonian forest, blanket bog, acidic alpine and montane grassland, many of which are protected through national designations such as Sites of Special Scientific Interest (SSSI) and Special Areas of Conservation (SAC). While Lynbreck is located within the park boundary, it sits on the outward fringes of the designated areas but it made us conscious that many of these habitats were present nearby.

Diana's report provided a list of the flora that she was able to identify as well as an insight into the many old ways the land had been historically managed. She divided the croft into four vegetation communities, groups of plants that grow in similar environmental conditions. In the woodlands, the smallest of the four vegetation communities, she discovered remnants of Caledonian pine woodland as well as ancient oak and birch woodland in a deep gorge known locally as the Slochd, an old Gaelic term meaning pit or pass. This was a real surprise as we knew there were plenty of birch around but we had never found any oak, only later understanding that while the tree itself was absent, the other species present in the general habitat suggested that oak may have been grown here in the past or even could thrive here in the future.

We referred to the Slochd as the gully, a steep-sided geological feature marking the boundary between our field and hill ground, carved by ancient processes when the land would have been covered in water, ice and retreating glaciers. The ground flora was besieged by the hungry mouths of our resident rabbit and deer population, but the hidden colourful sprinkles of wood sorrel, cowslip, dog violet and heath spotted orchid gave an indication of the potential for a carpet of woodland wildflowers, if only they could have the chance to grow.

In the rest of the woodlands, the main species of trees present included downy birch with the occasional alder, rowan, grey willow and eared willow. Diana noted the lack of other usually associated species, which she commented 'may be due to the history of grazing in these woods'. This was an important reminder of the long-term damage that repeated animal impact might have on what could be naturally species-rich and diverse habitats.

Diana described the grasslands, the third largest vegetation community, as being 'derived from past agricultural management', an indication that there was a lower diversity of plant species than would be found naturally. But she did discover a few areas where there was more diversity than we had initially realised, listing grasses such as cocksfoot, sweet vernal grass, sheep's fescue and Yorkshire fog, as well as wildflowers such as white clover, mountain pansy, tormentil and bird's-foot trefoil dotted between large patches of dry moss.

The habitat identified as mire, which is simply bog or boggy wet heathy ground, is the second largest vegetation community, the majority of which blankets an area we call the flats – around thirty-eight acres of gently undulating ground composed of a mosaic of misshapen bog pools, cushions of sphagnum moss and thick tussocky purple moor-grass. In the far southern corner is a drier, slightly raised circle of common heather with a ring of Scots pine trees around the periphery creating a natural grove in a place we call Bog Island.

With her experienced eye, Diana was able to identify 'enriched' areas, which possibly came about from past burning of the heather, depressions with no flowing water that were likely formed by peat cutting and elongated ditches of water, which suggested earlier attempts at draining the land. In between the visual storybook of previous human intervention, she talked of pockets where peat-forming vegetation was still intact and introduced us to aptly named wetland flora such as bog pondweed and bulbous rush.

The flats are also scattered with old woody stumps, which protrude like the exposed graves of giants, that once stood proud as mighty trees but, now sitting squat but upright on steady feet of old root plates, their physical remains mostly gone but their tree spirits ever present. To us, they tell a powerful story of a landscape that is no more, of a time when much of Scotland was cloaked in a rich diverse forest, where lynx and wolves hunted, wild cattle roamed and humans blended in to a landscape in which their impact was in much greater harmony with all other forms of life. Over a few thousand years, the dominance of the native forest abated as the climate changed and the rise of the humans led to deforestation for timber and the creation of farmland, leaving just isolated fragments of this mighty forest remaining. But our own observations of the landscape around us suggested the tide of trees was starting to return. Observing the slow ripple of pines regenerating towards us from the great forest of Abernethy, and the random pop-up groves like that on our very own Bog Island telling us that one day, if given the chance, the ancient trees that stood on the flats might just return.

The final and largest vegetation community that Diana discussed in her survey was dry heath, an area that takes in most of the hill ground and parts of the fields. The main plant that dominates in these areas is common heather, an iconic Scottish shrub, though of course not exclusively Scottish, with a deep purple flower and enchanting aroma that carpets the hills and woodlands of the Highlands in August. These are also areas with large numbers of naturally regenerating trees including Scots pine, downy birch, juniper and eared willow. On our wanderings, we noticed many stunted miniature rowans, which were clinging on to life despite the repeated browsing of any new growth by resident roe deer. Their fresh shoots make for tasty fodder as time after time they have their lead and side shoots eaten, only to fight back with spirit, but then to be defeated once again until, finally, the energy reserves in their roots dry up as the plant capitulates and dies.

On our western hillside beneath the regenerating trees is a row that runs north to south of around eight old stone grouse butts, man-made hollows where shooters would lie in wait for the red grouse flushed out to fly overhead, unaware of the camouflaged hunters beneath. There is a certain beauty about these old structures that hide in the landscape but we have learned to approach them with caution, their long heathery fringes convincingly concealing the deep carved earthen hole beneath, a murky depth that we have nearly tumbled into a number of times. Upon closer inspection, the sunken sides had been reinforced with drystone walling, harking back to a time of real craftsmanship when the human touch had been much lighter on the land. Today, most modern grouse butts are simple L-shaped wooden platforms that sit above ground, their sharp edges and perfectly parallel sides arranged in straight lines visually jarring on the dancing curves of the very land on which they rest and the backdrop of rolling hills against which they sit.

Diana noted some areas of dry heath, where the soils were thin with protruding rocks, as being of international importance due to the presence of a rare plant known as intermediate wintergreen. This was particularly interesting as the signs so far indicated that the land around had been, just a few centuries before, likely cloaked in woodland and was certainly reverting back to that now. It's often imagined that thousands of years ago, large swathes of land would have been covered in old forests but in many cases, the landscape would have been made up of a mosaic of new trees, older woodland and open spaces. This discovery of Diana's made us wonder if that was what our land at Lynbreck might once have looked like and so be worth taking into consideration in our future planning where our impact would change the land once again.

She also mentioned a track on our eastern hill where historical damage had been caused by vehicles driving up and down to the grouse butts on the neighbouring grouse moor. In addition to these observations, we had noted patches of shorter heather, likely burned in the past as a moorland management practice to get rid of lankier, mature heather and allow regrowth, the young shoots being ideal food for the resident red grouse. And, on the lower slopes of the hill, a series of small lumps and bumps, which caused miniature rises and falls to the thick blanket of vegetation, turned out to be piles of rocks called cairns. These were prehistoric field clearance cairns comprised of various sizes of boulders that had been excavated from the newly created fields and stacked into heaps to create 'productive' agricultural farmland. Today the eastern hill is a vibrant, young, regenerating woodland where the land has been rested from grazing animals, allowing thousands of stems of pine and birch to shoot up from the ground, poking through the rested heather. But we couldn't help but wonder if the ancient residents of Lynbreck played a role in the final clearances of the great forest that once stood, removing the land from the wilder clutches of nature.

Thanks to Diana's findings and our own improving identification skills, our relationship with the floral community was beginning to blossom, but we were yet to become properly acquainted with the extended croft residents: the birds, mammals and insects. Using our limited knowledge and a stash of books from our apprentice ranger days, we began to keep a diary of everything we could identify that flew past, buzzed around or scurried along. So many of these can provide a 'health check' to the condition of nature, a measure of the strength of its pulse. The simple rule is: the more you see and hear both in numbers of species and numbers of individuals, the greater the well-being of the land, and it's fair to say, we were kept busy.

In spring the sounds of curlew, a large mottled brown wading bird with a long downward curving bill, could be heard from out on the flats and was the first notable sound to awaken Lynbreck from its winter slumber. The name curlew comes from the phonetics of the sound it makes; however, to us it was a long 'towee towee towee' followed by a rapid 'churr churr churr', over and over again from first light to dusk. As spring is pre-midge season in the Highlands, we could stand outside on still evenings and listen to the snipe drumming in the wetlands as our resident woodcock flew back and forth, calling out a series of characteristic honks followed by a squeak, a part of its breeding display flight known as roding. And when the cuckoo arrived in May, the springtime orchestra of Lynbreck was complete.

It's astonishing how 'normal' it can become to hear all of these sounds and it's equally astonishing that the majority of these birds are at best 'rare', and in many cases 'at risk' as the consequences of the actions of humans – development, shooting and intensification of farming – has squeezed many of them into tiny islands of safety. It made us conspicuously aware of the significant uniqueness of the landholding we had become guardians of, where at Lynbreck, and in parts of the surrounding landscape, these birds were finding solace and solitude.

In early summer, the swallows arrived and took up residency in our old tumble-down stone byre. We'd watch the house martins dance over the fields, hunting for insects and listen to the red grouse 'yak, yayayayayayaya, yak, yak, yak' from beneath the regenerating Scots pine on the hill. On still, calm, warm days there would be a vibration of butterflies above the grasses, wildflowers and shrubs in our fields, woodland and bog edges. Early on, it was the orange-tips, small tortoiseshell and peacocks, followed by the common blue and small pearl-bordered fritillary and finally the Scotch argus taking us into the final days of summer. By early October the migrant birds had left and the butterflies gone as the air became filled with the distant groaning of the red deer in the forest, a time known as the rut, when males establish their territories, calling the surrounding females to gather for mating.

By early autumn of our first year, we had a good understanding of the lie of the land and were beginning to get to grips with our new roles as stewards of a diverse, mixed-habitat landholding that was sharing its own story. For centuries the croft had been altered by people, a timeline that could be traced from deforestation to settlement, sport and agriculture. Our thoughts were often revisited by the images of the old tree stumps that littered the lower parts of the croft as the spirits of the Lynbreck Ents, a nod to Tolkein's tree-like beings in the *Lord of the Rings* trilogy, haunted us from day one.

Rather than just reaching for books or asking experts, we looked to the land in the first instance as we tried to find our place and understand our purpose. Alongside our plans for a new farming business, our goal was also to define our own role, one where our presence or actions would not take away from the life that was already here, but add to it instead. It also wasn't about returning the land to some former idealised state, it was about reconnecting and re-harmonising the hand of humans with the hand of nature, the two clasped together on a new path at the very beginning of our journey into farming with nature.

On that first day as new owners, our inaugural reconnaissance walk took us towards and up the western hill ground of Beinn an Fhudair. As we began our ascent along the old stone farm track, a route the old crofters used to access the higher peat basin for cutting fuel, scattered all around us were Scots pine and birch seedlings, some ankle height, some metres tall, and accompanied by the odd clump of eared willow and an occasional hidden juniper. Like us on that walk, they too were climbing, heading for the summit, and judging by the different ages, had been for some time.

There had been a fair bit of nibbling from the local roe deer population, stunting the rowans in particular, but overall the numbers of deer were low, enabling some of the other trees to 'get away' and establish. After an hour or two we made our way back towards the homestead taking a different path, this time descending down the steep slope into the gully where it felt as if we had entered another world in which a few scattered ancient trees stood strong. Its beauty was so striking and its energy so apparent that a friend described it as a magical place where fairies would live, if you believe in such things.

There were a pair of giant aspen trees, a relative rarity in Scotland but one which we knew from our tree planting work in the Borders should be more widespread, as well as multi-stemmed stumps of hazel so substantial that even if we joined hands, we couldn't wrap our arms around all the stems in one go. There were birch with trunks bigger than we had ever seen, and old rowan and juniper that clung to the craggy rock faces above. But while the beauty of this hidden hideaway was distracting, we instinctively knew there was something crucial missing and recalled an experience from just a few months before.

CHAPTER 7

The Great Lynbreck Tree Plant

We'd taken a couple of days to head north from the borders to visit Alan Watson Featherstone, the founder of the charity Trees for Life. Alan had been a bit of a hero of ours and had recently become a friend following a group study tour to Norway to look at natural forest regeneration that I was fortunate to go on with my job of that time.

We spent the weekend at his house in the ecovillage community at Findhorn on the Moray coast and on the Sunday he offered to show us some of his early work in Glen Affric, a large estate to the south-west of Inverness where ancient trees blend into treeless moorland tops, broken only by a patchwork of lochs. On first impressions, it's hard to not be impressed by Glen Affric as the ancient Scots pines, known affectionately as 'granny pines', stand tall like centuries-old cathedrals, built with resilience and gnarly precision and where the spirit of place is eminently present.

As we took a short walk into the woods from the car park, Alan helped open our eyes to what was really happening around us. The trees were magnificent, aged and wise but slowly they were slipping into a childless senescence where there were no saplings, young whips or semi-mature stems growing up from the annual rainfall of seeds. Nothing. It felt like a waiting room for death with no sign of new life. Alan explained that the new seedlings had no chance to gain strength and grow because of the high numbers of deer browsing on them.

Our walk continued up a small incline through the forest to a deer fence as he told us how it felt when the realisation of what was happening hit, compelling him to take action. He understood that in order to let the trees establish, they had to be kept safe from foraging herbivores and the easiest, quickest and most effective way to do this was to find a suitable area and put a big fence around it.

The results, a few decades on, spoke for themselves as we stepped through a gate into a world of vibrant new life where Scots pine seedlings were popping up everywhere. Nature had finally been given the chance to work, placing the right tree in the right spot and creating a succession of new trees from old stock. These granny pines were finally allowed to pass on their genes and they would now see out their days accompanied by the next generation.

That weekend gave us much to think about after a series of intense but enlightening conversations talking trees, rewilding, food choices and our impact on the planet. We lived and breathed the spiritual community of Findhorn for forty-eight hours, feeling a strong connection to a people who had the sacredness of nature at their core. We didn't realise it then, but on reflection it was a micro moment of time in our lives that had a significant impact on our future actions. Standing in our ancient wooded gully now, those emotions from Glen Affric revisited us and, just as we came away from there with hope, we had a similar feeling once again. It was as if the clouds parted to reveal a path, where the land's call to action was heard. We would protect what was left and we would plant what was missing.

For us, planting a woodland was a personal dream, stemming from our love of trees that had lain dormant in us for many years. As a young woman, Sandra's dream had been to become a forester but she was deterred by the fact that it was more something a man would do and the career advice given directed her instead towards training to become a librarian. When I was sixteen, I explained to our school careers advisor that I had a passion for history, nature and the outdoors, wanting to somehow work within those fields. I was sent on a two-day placement with the National Trust, shadowing office and shop staff, and came away from the experience assuming that maybe what I actually wanted to do didn't exist as a career option.

During our time as tree planters in the south of Scotland, we missed the company of our old tree friends as we spent day after day planting the next generation in empty hills where the ancient Ents of those landscapes were no more. On some days the air hung heavy in the empty valleys and you could feel the weight of the sadness of the land that had repeatedly suffered for human gain, where the bountiful ancient forests had been removed and later replaced with sheep. It was a landscape we never bonded with, never really felt comfortable in, acutely aware of the lack of diversity there. But Lynbreck was different. Here, some small trees were able to return of their own free will and we also reconnected with our ancient tree friends who, just like at Glen Affric, were in need of assistance to successfully reproduce. Diana's research had opened our eyes to a landscape of the past and a landscape of the present, giving us the knowledge we needed to plan for a landscape of the future. One where a connected patchwork of mixed, native woodlands would be full of life, where nature could spread her wings and multiply, and where our team of animals would only benefit and add to a more diverse and multi-layered landscape.

We decided to make contact with the Woodland Trust, the Cairngorms National Park Authority and Scottish Forestry for advice and a second opinion on our thoughts. It was essential that we were doing the right thing for the land and even though the conversations highlighted some potential challenges ahead, such as where exactly to put a protective fence and how best to plant the trees, the overwhelming feedback was very positive and encouraging. The plan was to access a pot of funding provided by the Scottish Government for new woodland creation and, as our research began on which trees would be the most appropriate to plant, the report from our vegetation survey came just at the right time. Diana talked of the damage that herbivores, in this case mostly rabbits, were doing in the gully woodland and highlighting the need for their exclusion if we wanted to give our ancient trees a chance to successfully reproduce. Further investigations using the online Scottish Forestry database indicated that parts of the western hill ground area that we were proposing to plant would be suitable for an upland oak woodland, which would include species such as eared willow, grey willow, alder, downy birch, rowan, hazel, hawthorn, bird cherry, wych elm and holly. If these had existed before, many of them were now absent, with a few exceptions found in tiny corners on the croft or in the wider surrounding landscape.

By October 2016, just six months after arriving at Lynbreck and having just submitted our application for the Young Farmers Start-Up grant to get support for the proposed new farming business, our woodland funding application was approved. But, while this was a major milestone reached, there was yet another barrier that stood in our way. The rules of the grant were such that we had to pay any costs up front and then recoup them back through a claims process, a hurdle that we later learned is one of the major reasons why smaller-scale planting projects like ours fail as the upfront capital is simply not available. The fencing alone was going to come in at around twenty thousand pounds and that had to be installed to create a protected area before we could start planting.

During our time in the Borders, we had seen planting projects where the charity I worked for, as well as private landowners, claimed something called carbon money to help bridge the gap in funding. This was when the amount of carbon a new woodland could draw down over a given period of time would be given a monetary value and turned into a tradeable commodity for sale to companies or organisations that want to offset their emissions. The process is usually handled by a carbon broker who, after taking a percentage themselves, passes on the rest of the money to the landowner to help fund the project.

We learned that there was a good chance for accessing nearly enough funding to cover the costs of our fencing but, like the planting grant, it would only come after works were completed. Determined to find a way through, I picked up the phone and dialled the number of the company we had been in conversation with called Forest Carbon, who specialised in this area, explaining quite honestly our situation and that, in spite of their offer of funding, we were no further ahead in making progress on the ground. In an unexpected twist, the director, James, whom I had met a handful of times previously, offered to give us all of the money upfront, drawing on our reputations as trustworthy and hardworking, pledging his faith and their money towards our new venture.

The market place of carbon trading has since boomed, particularly in relation to woodlands, but now includes peatlands and even farmland. It follows a 'polluter pays' model where environmentally damaging practices can be 'offset' by buying the work of someone else. Many are critical, arguing that it is not the solution to addressing the challenge of climate change, whereas others see it as an incentive to drive positive changes in land use. Either way, without it our project would have been dead in the water and we were very grateful for the generosity shown to us by a virtual stranger who had been inspired by the bigger picture of our Lynbreck vision.

As the excitement and anticipation of what lay ahead began to build, we started to feel a little intimidated by the size and financial value of the project we had taken on. By signing contracts with Scottish Forestry and Forest Carbon, we were now legally obligated to ensure that the planting would be successful. If, for example, the fence was breached after a night of heavy winds or deep snow, and flooded by hungry, tree-foraging deer, hare and rabbits, our work could be undone in a matter of hours. The financial responsibility would sit with us to replant and if the problem continued to happen, the project could be deemed unviable, putting us in a situation where the investment would have to be repaid, a prospect which we couldn't fathom.

Determined to keep going, the deer fence was completed by the new year and we became busy with preparing the ground for planting. With 17,400 trees to plant, we had to clear the vegetation from 17,400 random spots so as to knock back the chance of natural competition and giving the young trees a head start. In many situations this can be done by a digger or by spraying a herbicide, a chemical application that kills off any vegetation it comes in contact with. Neither of these were suitable methods for us as we had concerns about the implications of ground compaction that a digger would cause and we had a zero-use policy on any chemical applications.

The majority of planting spots would be in varying heights of heather. As a slow-growing shrub, our thinking was that if we could reduce the height down to ground level and directly plant into these bare spots, covered only by a natural mulch mat of moss, we could be recreating, in as natural a sense as we could make it, the perfect conditions for a tree to establish and grow. And so, every day for two months, one of us would climb the hill with a brushcutter – a large strimmer with a spinning metal head – and cut over seventeen thousand square patches at two to four metre random intervals into the heather on our hillside, while the other would follow marking the spots with small canes. There were a few grassy patches to be planted into and in those we used a mattock to carve off the top layer of vegetation and expose the bare soil beneath.

By mid February, the spots were ready and on a wet, grey day at the end of that month, the trees arrived. Bag after bag of mini trees were offloaded at the entrance to the old croft house where we would store them until they were each taken onto the hill for planting. Our contract stated that the project had to be completed and the claim form submitted by 31 March, leaving us with just over four weeks to plant all of the trees between the two of us. Some thought we were mad, others that it would be impossible and that we had been naïve in our plans. But it was the part of the whole project that we felt the least anxious about, aware that it would be physically demanding and exhausting, but more than ready for the weeks that lay ahead.

And so, every day at dawn, we headed onto the hill with bags weighed down with small trees and planted until dusk. Some days it might only be a few hundred trees each if the terrain was particularly challenging and on other days nearly five hundred on the flatter sections. The winds battered us, the rains soaked us, and the snow and icy blasts chilled us to our cores but, onc by one, the trees went into the ground. Every day we would tally up the latest count and while progress was being made, there was no opportunity to stop even as physical exhaustion began to seep into our very bones. Some days just lifting the planting spear felt like a lot of effort as arms became heavier and legs felt weaker under the weight of a newly refilled planting bag with a slope to climb. One day, I only managed to plant until lunchtime, spending half of the morning weeping as I planted, feeling so utterly exhausted, pathetic and sorry for myself that Sandra sent me home to bed as she carried on.

But on those days when the sun did appear, the brightness and warmth of the rays recharged our bodies and spirits. At lunchtime, we would sit in the heather propped up against the quad bike, eating a lunch of oat cakes and homemade biscuits while sharing a flask of tea. Those were the times when the tiredness abated, the enormity of the task disappeared and we felt so happy and so lucky to be there, in that moment. And, after one of our most challenging days, when we planted until dark as progress had been slow due to sporadic snowfall making the steep slopes we were planting on treacherous to navigate, we treated ourselves to takeaway chips. It's not something we did often but, on that day, every deep-fried, guilt-free chip consumed was the best thing we had ever eaten and it was these little moments that just lifted our spirits and helped us to keep on going.

It's a very powerful experience to plant a tree and we were all too conscious of the consequences of our actions, fundamentally changing this landscape. Our days planting required not only physical effort, but mental focus. We would carefully select the right spot for the right tree so that species such as alder and willow would be planted in the wetter areas, keeping species such as hawthorn and rowan for the drier ground. On the bracken-dominated slopes where the soils were richer, we would plant oak and hazel. Most of the tree whips were only thirty to forty centimetres but would grow to be many metres in height. These saplings would become giants that would tower above the croft. While taking only a minute to plant, they would stand for many years, their roots penetrating deep into the earth and their branches stretching outwards into the open air. We couldn't help but get philosophical about the whole experience, acutely aware of how much Lynbreck would change in our lifetimes.

On some days, we wondered what the locals must have made of all of this as they watched us day in, day out, trundling up the hill throughout the winter and now finally planting. We later found out that our neighbour George used to watch us through his binoculars, intrigued to see our progress. His friend John used to call us the tree-planting bumblebees as our dark clothing and white bags with yellow straps made us look like giant pollinators buzzing around on the hill. In some ways, that's exactly what we were.

And, as we kept going, the number of trees still to plant began to be fewer than those already planted, and during the last few days our energy levels rose a little as we gave our efforts a final push and ended up finishing just a few days before the deadline. It was a moment of huge relief marred with a slight twinge of sadness. This was very probably the first and last time we would ever get to plant our very own forest and now it was done. As a rare but well-earned treat, we took ourselves out for lunch to our favourite café in Findhorn, just an hour north on the Moray coastline. Once again, I was battling extreme exhaustion and, after we returned home, I spent the next three days in bed, every part of my being literally crying out for rest and recuperation. In contrast, Sandra carried on with other croft chores, her body impressively unfazed by the four-week marathon as I lay in a crumpled heap. And, while physical recovery was one thing, mentally we didn't have time to stop for too long. With one project done, it would soon be time to start the next, that for us would be a much greater challenge. With our Young Farmers Start-Up application approved just a couple of weeks after we finished planting our fledgling woodland, it was time to put the wheels in motion of our new business. We had the plan but now we needed to figure out the detail. How could we make our farming enterprise deliver for nature? We knew what we wanted to achieve but was it really possible? And, if so, how? There were many questions still to answer, many details to still add before we could really get this plan off the ground.

CHAPTER 8

Becoming Farmers

During our apprenticeship training with the National Trust, one of our college lecturers introduced us to a documentary called *Natural World: A Farm for the Future*, which was written and presented by Rebecca Hosking, a journalist who had grown up on her parents' farm in Devon. Rebecca highlighted the pressing urgency to farm in sync with nature for the health of our planet, emphasising the importance of connecting people with where their food comes from. There was one section in the documentary where she took apart a sandwich she had bought at a petrol station, deconstructing the physical contents and identifying each element and the amount of energy it took to produce. Rebecca broke down what on the surface looked like an innocent quick lunchtime bite to eat into an energy-hungry, intensively farmed, chemically treated consumable product. It made compelling viewing with a lasting impression.

The questions we found ourselves asking is what does farming in harmony with nature look like? And is it even possible without having some sort of negative impact somewhere down the line? We had ideas of avoiding chemical use, working with traditional and native breeds of livestock, nurturing fields full of grasses and wildflowers, planting trees and selling our produce to our local community. Our land would hum with life in all seasons as we reaped the bounty of food that our partner nature would provide. It was all so clear in our minds and sounded so perfect and idealistic but was it actually realistic? And, if it made so much sense, was so bountiful and harmonious that it was not just good for nature but could help mitigate the climate emergency, then why weren't all farmers already doing it?

And so our search began to find others in Scotland who were farming in a way that Rebecca had spoken of and that matched the ideal in our minds. An obvious starting point was to look within our surrounding area, acknowledging that the landscape we lived in provided its own unique challenges. I remember a locally born and bred farmer from just a few miles away, lower down the valley, saying he believed that our location was one of the most challenging conditions in the whole country to farm. Our high altitude would bring snow in the winter, making access to and around the croft at best challenging, at worst impossible. The exposure to the prevailing weather systems from the south would see us battered in all seasons with storm force winds so powerful they once blew an empty water tank the size of a small car across our field in August and perilously in the direction of our wooden cabin, as if it were nothing greater than a bundle of tumbleweed. Through experience we learned that if things weren't cemented down, tied down or held together with a shedload of nails, they would be shattered into pieces and scattered to every corner of the croft. And the heavily acidic, wet, peaty soils that are so typical of the Highlands of Scotland meant that land fertility was low, providing yet another natural limiting factor on exactly what we could grow and produce at scale, especially if we were to avoid artificially altering the soil with chemical inputs and fertilisers.

We looked to friends and neighbours for advice and practical experience on working with animals and other useful tips; we read magazines and books, went to agricultural shows, talks and gatherings, trying to glean as much knowledge and information as we could. The more we became exposed to mainstream farming, the clearer it became to us just how extensive the agricultural industry is. If you go to any country show or open an industry magazine, it is full of organisations and company adverts selling seed, fertiliser, herbicides, pesticides, fungicides, lime, harvesters, tractors, feed and minerals, all of which come with a promise of increased productivity and greater financial return. A picture had started to unfold, one where there was a lot of money changing hands, the bulk often leaving the farmers' bank accounts, so that they could farm in a certain way, one that it seemed the corporate industry – sometimes referred to as 'Big Ag', short for big agriculture – and the government, was driving.

It's not uncommon for farms today to run at a loss with many at risk of dissolution should the subsidies be discontinued, and it is easy to see why. There is overwhelming pressure to buy the newest kit (that promises greater efficiency), sow the latest seed (to give greater yields), apply the improved fertiliser (to get faster crop growth) and purchase or breed the biggest fastest-growing animals (to achieve the heaviest final weight in the shortest amount of time). It seemed that in order to be the best farmer, you have to pay for it. But, we wondered, at what cost? The kit would need upgrading every time technology advanced. The seed would need to be bought again. The fertiliser would have to be reapplied in greater quantities as the natural energy system of the soil would gradually deplete. And those big animals usually require a huge amount of bought-in feed to bulk them up and often come with their own biological problems, such as requiring assistance during birthing. I remember hearing a story of a local farmer that was about to go bust. The night before he was due to lose the lot, he was still saying, 'If I could just buy more animals, I could make more money,' when the reasons he had got into difficulties were that the desire to grow and produce more always resulted in greater expense over income.

The cloak of the mantra 'we have to feed the world, we must produce more' hides a darker shadow, sweeping aside the voices promoting farming with nature or organic farming as unrealistic and with limited capacity to produce, both of which are untrue. It has always felt unfair that the weight of 'feeding the world' sits on the shoulders of farmers trying to make a day-to day-living for their families when the pressure to produce more can often lead to greater annual financial deficit. It was becoming clear that more income and more output did not usually mean more profit, and it was a rabbit hole that many had fallen down that we wanted to avoid. We preferred to see 'feeding the world' as growing food to feed ourselves and then using the rest we could produce to feed as many in our community as we could. This was our real world, our day-to-day world – one that we could relate to and would have a direct and measurable impact on.

It has become all too easy to launch attacks at farmers without actually understanding how the system works. Government subsidies and Big Ag have snared farmers into a dependency on subsidies and inputs, and it's become a system so convoluted and complicated that it can be hard to see a way out. Many will work to the best of their ability within a framework that dictates the kind of farming that will be rewarded financially. But, by buying into that system (and to not do so, or to leave the system once in it, is incredibly difficult), the farmer forfeits many of their opportunities to run their business as an independent set-up.

In order to get the most annual government subsidies through the largest of all the schemes known as the Basic Payment Scheme (which comes with its own long list of eligibility requirements), farmers must make their land as agriculturally productive as they can, requiring them to apply a whole host of inputs, whether mechanical, chemical or feed related. When the livestock or crops are ready to be sold, their true value – the true cost to produce them – is rarely reflected in the price the farmers are paid, and this is where the subsidies kick in. Farming subsidies don't exist to prop up a farmer, they exist to subsidise the actual cost of producing food to make it cheaper for all of us.

In addition to these external pressures, by signing up to government support schemes, the farmer is subjected to regular inspections that are often unannounced or with twenty-four hours' notice, where any kind of violation or slight deviation, no matter what the reason, can result in a penalty, a word in itself that implies punishment. But it's a catch-22 situation as without these subsidies and schemes many would not survive. An accountant friend estimated that around 90 per cent of their farm clients ran their businesses at a loss, an alarming statistic which, if accurate and indicative across the board, illustrates just how troubled the situation is for the people on the ground. Farmers are often judged very quickly and harshly by their critics, accused of endless crimes of water pollution, soil degradation and wildlife persecution, but understanding the context that can drive these practices suggests the hand of responsibility sits much higher up the chain.

For many, turning their backs, walking away from the financial support that keeps them afloat and the modern culture they are familiar with, is a price too big to pay and an option too frightening to contemplate. And when the money is available to them, why should they? This can lead to an entrenchment of ideas and practices on opposing sides as the battle lines are drawn with those who threaten them. With mental health issues and suicide rates on the rise within the farming culture, we began to realise how important it is to understand why things are the way they are, with open hearts and minds and before the first grenade of criticism is thrown or bullet of disparagement is fired.

This stark realisation led to a turning point in our lives, acknowledging that perhaps our inexperience in farming was a blessing in disguise. It allowed us to see the realities from the outside, free of cultural expectations and annual subsidy shackles, and enabling us to find our own independent route forward. We had not been exposed in the same way as those growing up or being educated in farming to a continuous stream of 'this is how you do it' or 'this is how it's always been done' or 'this is what you need' or even 'this is how much money you could (or should) make'. Nor did we see land as a commodity where the rhetoric between 'farming' and 'nature' reinforced a relationship akin to master and slave.

In contrast, we had started to see good farming as an art that relies on a close relationship with the land worked and the animals raised or the crops grown, and we found a small but bright beacon of light just two hours away that gave us hope of a different way of farming.

Roger and Gilly are a pioneering couple who run the Sailean Project on the Isle of Lismore off the west coast of Scotland. Lismore has a warmer, wetter climate and is formed on an elevated rocky outcrop of predominantly limestone, resulting in much more fertile soils. So, while the conditions there are actually quite different to Lynbreck, it was the closest working example in terms of farming approach and practices that we could find of what we were aspiring to do.

When we visited them at an open day in the summer of 2017, we learned that Roger used to be a large-scale arable farmer in Cambridgeshire, spending many years tilling the soil, dousing the land in chemicals to grow high-yielding crops that would deliver the biggest financial return, and, he'll tell you himself, he did it very well and very successfully. After moving to Scotland to retire, the couple had no intention of getting back into farming but found themselves with land at Sailean and space for some hens, cattle and sheep. Roger had also watched the documentary by Rebecca Hosking, something that caused him to face a crushing realisation of the damage he had done by farming an arable unit so intensively for many decades. From that point on, they committed their lives to running a new, diversified farm business based on holistic principles that would turn a profit, regenerate the soil and biodiversity, produce nutrient-dense food and educate others. We felt we had found kindred spirits in Roger and Gilly and, while their situation was very different, it was so exciting to find a real live working example in Scotland of the kind of farming we wanted to do.

Around the same time, we learned of a relatively new movement taking hold of the alternative agriculture world that was referred to as regenerative agriculture, a way of farming that regenerates the land and the communities that live on it. We started to read books by some of the leading figures in this new movement who, while critical of the industry to date, spoke with such passion and positivity about a different way of farming where everyone could benefit, if only we worked more with nature. There is Joel Salatin from Polyface Farm in Virginia in the US, an entrepreneurial and hugely successful farmer with a big personality who talks openly about the state of food and farming today and is blessed with a way of communicating that presents complex science in layman's terms using an engaging and accessible format. There is Richard Perkins from Ridgedale Farm who had bought a small landholding in Sweden and within a few years transformed it into a highly efficient permaculture food production unit with a huge following and influence after sharing educational videos on his YouTube channel and running on-farm courses. And then there is Allan Savory, a Zimbabwe national who became famous for his TED talk 'How to Fight Desertification and Reverse Climate Change' and talks passionately about his solution for regenerating landscapes through a framework he calls 'holistic management'.

While they each had their different approaches and individual styles, they all talked of a world of farming that focuses on the importance of soil health, looking in particular at soil biology and working with livestock to help build even more soil and biodiversity, all embedded within a healthy, functioning social community. In our early days at Lynbreck, we attended a few farming events where soil and soil health were discussed at length but, according to the many experts we heard, true soil health would only be attainable with often substantial investment. For many decades, the farming industry has focused on the chemistry of soil. Nitrogen, phosphorous and potassium, often referred to as NPK for short, are seen as the most important elements that need to be added to the soil regularly in the form of synthetic or natural fertiliser to keep the soil healthy and vegetation productive. At these talks, we were told that soils can be lacking in certain minerals so animals would need to be given these as supplements to avoid any deficiencies. The financial outlay for all of this was certainly starting to add up and we found ourselves questioning how soil, that forms the building blocks of all life, can be naturally deficient in anything? Or is it the repeated, intensive, invasive impact of humans that has caused the problems that we were now trying to cover with a technological sticking plaster, but not actually dealing with the deep wound beneath?

We would come home from these talks with our heads spinning, feeling bamboozled with the complexity and costs of what we were presented with. Sandra recalled a talk she had attended just after we moved to Lynbreck in 2016 by a lady called Christine Jones, an Australian soil ecologist who had talked at length about soil health, but with an emphasis on soil biology, not soil chemistry. In one of her many published papers, called 'Light Farming', she explains in simple language, the importance of harvesting the power of photosynthesis to build healthy soils and increase the profitability of a farm business, which did not require costly artificial inputs or supplements.

Christine explains that pasture farmers are actually solar farmers who capture sunlight via the leaves of the plants they grow and through the biological process of photosynthesis. This solar energy is turned into sugars, which are the building blocks of life. The sugars are then transformed into a number of carbon compounds, which are essential for creating topsoil and support a vast community of soil microbes. When there is a healthy microbial community that is connected to plants via a superhighway of underground fungal networks, plants are able to access the minerals and trace elements they need and these are, in turn, passed on to the animals that graze or browse them. And the way in which to create or maintain an optimum functioning soil biome is very simple: keep the soil covered, grow a diverse range of flora, avoid chemicals and integrate animals into the system. In her paper, Christine is essentially giving us the choice to either regenerate or degenerate the land based on the decisions we make and the approach we take in our farming operations, explaining quite explicitly that soils are not deficient in anything if their biology is working at optimum level, and that is what we had to focus on.

The more we read, watched and listened, the more our ideal of what a farmed Lynbreck landscape would look like took shape. Our business plan had already been approved but we were really starting to learn how we could turn a largely aspirational document into a real working model. The voices and working examples were few, but they were loud, powerful, compelling and inspiring, and we realised that if the regenerative path was for us, it was one we would have to carve out for ourselves. And while exciting, it was a prospect that initially felt overwhelmingly intimidating as our approach would take us against the advice of seasoned farmers, shunning the rhetoric of mainstream agricultural education and essentially going it alone.

On one particular evening when doubts were gathering, as they so often did in those early setting up days, Sandra called Roger and Gilly for some advice, searching for the reassurance we were needing. In between lots of useful advice and encouraging words, Roger said, 'Work with what you've got,' a phrase that we revisit and repeat to ourselves to this day, adopting it as a baseline motto when we consider starting new ventures or trialling fresh approaches.

After the call, we began to brainstorm 'what we had', mind mapping a page of everything we could think of. We took a sheet of A4 paper and wrote 'Lynbreck' in the middle, drawing lines from the centre to key words and sentences: grasslands, woodlands, hill ground, bog, existing and planned buildings and infrastructure, personal skills and knowledge, community, our location. Soon the page was filled with a collection of physical, natural and cultural assets that combined gave us all the ingredients we needed for this new farming business.

Roger and Gilly had also inspired us to delve further into the details of holistic management, a decision-making framework based on delivering a triple bottom line of positive environmental, social and economic outcomes, and the brainchild of Allan Savory, one of our early influencers. We were later given the opportunity to attend a three-day holistic management course, delivered by a farmer called Tony from Meeting Place Organic Farm in Ontario, Canada. The training started with Tony standing silently at the front of the room next to a flipchart on which was some handwritten text that said: 'You are the expert on your farm.' This short and simple, yet intensely powerful statement is not something that farmers are often told in the context of an industry dominated by advisors and salesmen. We spent the day learning how to write our own 'holistic context', a series of statements written in the present tense that set out how we want our lives to be and how we will enable it to happen. We wrote things like:

> ***Quality of Life (how we want our lives to be)***
> *We have a strong personal and professional relationship and have a sustainable work/life balance. We value our way of life and our way of farming. We nurture our physical and mental health. We are financially secure in the short and long term. We grow and raise the highest-quality produce on land free from artificial inputs and using ethical feed. Our animals have everything they*

need to stimulate their natural instincts and live a low-stress life. We provide healthy, nutrient-dense food for our local community. Our way of farming and living is designed to have a positive environmental impact. We encourage a regenerative way of farming and reconnect people with real food.

Form of Production (how we will enable it to happen)
We maintain good channels of communication. We take an annual holiday and have a weekly celebratory meal. We stay true to who we are and what we believe in and take time to enjoy our croft. We grow, source and eat ethical food and take regular exercise. We carefully plan our finances and know all of our costs and profits. We avoid unnecessary debt. We look after and build our soils. We respect and work with the animalness of the animal. We grow and raise as much food as our land can sustain and sell it within a small radius. We farm with nature. We make Lynbreck a holistic, educational hub.

This was quite a different experience to writing an aspirational document that focuses more on hopes and dreams. Instead this was underpinned by personal morals and values, which would allow us to fundamentally achieve fulfilment and happiness in life. The finished statement would act like a lighthouse that would guide us safely back to shore when the seas became choppy and unnavigable, a point of reference to help us continually review what we were doing and why we were doing it.

And, in clarifying who we are and what we wanted, we had started to question how we expressed ourselves to the outside

The house hens are completely free to range, often spending time sheltering and resting in our woodshed.

Hens are naturally inquisitive animals and love nothing more than a human perch.

Our flock is a mixture of pure and cross-breed layers who produce a rainbow of eggs including brown, white, blue and green.

Pigs are domesticated from wild boar and thrive in natural habitats.

Oxford Sandy and Blacks are known as the woodland pig.

We make time to socialise our pigs from a young age, helping them to get to know and trust us.

The flats between Lynbreck and the mountains were once a glacial lake.

A view towards the homestead of Lynbreck.

The breathtaking view towards Abernethy Forest and the Cairngorms, something we never tire of.

Highland cattle are native to this area and thrive on a range of natural vegetation.

The cattle are outdoors all year round, dealing with some harsh conditions including high winds, low temperatures and fierce blizzards.

Our livestock are a part of our team, where calm, respectful animal handling is the basis of our relationship.

Decaying wood provides habitat for a wide range of wildlife.

One of the nearly thirty thousand trees we've planted at Lynbreck.

Robins often shadow us as we work around the croft.

Outdoor cooking of Lynbreck produce over homemade charcoal is a favourite year-round pastime of ours.

We use the traditional craft of food smoking to flavour some of our artisan meat produce.

The kitchen garden sits at the heart of the Lynbreck homestead, providing food for us to harvest year round.

Our resident bee population works as pollinators, gathering nectar, which they turn into honey.

Each summer we harvest a small annual crop of wildflower and heather honey.

world and how the outside world might see us. Neither of us feel comfortable in defining ourselves with labels or attaching ourselves, our work and our ethos strictly to one movement, but we sometimes use such language to help explain what it is that we are trying to do. The labels become waymarkers that help us to navigate the different routes ahead, also marking where we have come from and, as a result, they become a part of the story of our journey as well as reference points for others who may follow. Words are an incredibly powerful tool, which, if said often enough in a certain context, can create or affirm thoughts, modify prejudices and influence mindsets.

We felt a strong cultural connection to the crofting tradition we had become a part of and often talked about working in crofting and being crofters. We would also refer to ourselves as farmers and working in farming, terms that were broader and more widely known outside of Scotland, covering everything from farming and crofting, growing and producing. But how might our vision of farming fit in with the other labels we had used in previous lives and our work in conservation as land managers? We had no intention of trying to 'manage' the land, a term which never sat right with us, implying as it did a human-dictated dominance over the land. To us, it's always seemed like quite an arrogant thing to think that we can manage – in essence control – nature, when in actual fact we just end up fighting it. There was something about conservation that felt awkward, too. When we try to conserve something, we choose to maintain a habitat or species as it is or was at a particular point in history. But to stop conserving could mean the loss of species which exist today, and so the rabbit hole becomes deeper and deeper.

The question in all of this revolves around what our role in this world is and what we are trying to achieve? To not manage land involves letting go, giving that control and responsibility back to nature and instead adopting an approach to work with it – a fundamental shift in practice and mindset. And we found ourselves asking how such a radical shift could fit a model of farming, one of the most impactful human actions that has shaped the modern world into what we know it as today? Our goal was to remove the focus from asserting our control over the land to managing *ourselves* and *our* actions instead, to play *our* part as biodiversity enhancers and enablers by assessing every action we take on our land on whether it would have a positive environmental outcome. What we were trying now to learn was how.

As we started to question all these mainstream labels and approaches, we found ourselves drawn even closer to the relatively new movement of rewilding; an approach that aims to use human intervention to help restore nature to a more harmonious and ecological balance within itself, and then reducing this intervention over time as natural systems begin to operate at a more optimal level. Rewilding is a topic that has grown in mainstream popularity but it has also developed a culture of polarity where the needs and wants of nature are often pitted against the needs and wants of people. Before Lynbreck, we spent two years working on rewilding projects and the experiences we had and the people we met along the way made a huge impact on us. We saw firsthand the transformative results that planting trees, removing old fences, restoring peat bogs and reducing grazing could have on a landscape formerly depleted in natural diversity.

But, as our lives started to transition into full-time farming, our active affiliation with the rewilding movement dwindled. Many individuals have their own definition of what rewilding actually means and the conversation is usually dominated by talk of nature in a *separate* sense to humans, often in a way that suggests it *does not include* humans and where the main topics generally focus on the reintroduction of species such as beavers, lynx and wolves. These are commonly referred to as keystone species – those that have a huge impact on the land they live on, often changing habitats and affecting densities of other wildlife. They are given major emphasis in the rewilding movement as they play such a fundamental role in redressing the ecological imbalance.

However, one of the problems is that these animals can also be seen as posing a direct threat to farming operations because their impact could have a fairly swift effect on the land and on livelihoods. The worry is that beavers, as they build their dams across rivers, might cause flooding of fields, and lynx and wolves could hunt farmed livestock. These perceived threats then dominate the debate as views become increasingly extreme and entrenched and the arguments become louder and more frenzied.

There are many in the rewilding movement who try to encourage a conversation that includes people, but the dominant, more extreme rhetoric can drown these voices out. The fact that we are killing off species and habitats is talked about a lot, but what we don't talk about enough is the role that people play as *a part of nature* in our roles as mammals, omnivores and predators, acknowledging that it is here that the problem lies. Nature, as a system, is not broken, it is just wounded by the continuous dominance that we inflict on it. Nature doesn't need to be rewilded, people do. As we came to understand this, our emphasis shifted from restoring landscapes and reintroducing species to reconnecting people to the land and rebuilding our relationship with nature to shape positive change collectively.

And, as we began to question all these labels and ways of working, we started to feel increasingly distant from all the communities which we partly belonged to – farming, conservation, rewilding. Neither of us have ever seen ourselves as pioneers or entrepreneurs, nor have we particularly looked for ways to stand out from the crowd or purposefully disrupt the mainstream. Yet, individually we are both quite headstrong, the kind of people whose heart and gut rule the head, using our instincts to guide us rather than what is trendy, popular or new. Gradually, we found ourselves feeling somewhat isolated, navigating through the new life we had chosen, but where the guides were few and far between. And so we eventually accepted that the closest approach we could find that best matched our aspirations was that of the growing movement of regenerative agriculture, which, in its purest form, is a way of producing food with an underlying current of 'regenerate everything as you go' – in essence our soils, our land, our people and our communities.

By the end of our second summer at Lynbreck, our vision was simple: to farm in a way where the impact of our animals would benefit the health of the soil and increase the diversity and abundance of species both below and above ground, and we reflected again on the observations of Allan Savory. For years, Allan had watched the African plains desertifying, resulting in a mass loss of wildlife, habitat and livelihoods. Initially, he placed the blame at the feet of wild herbivores repeatedly taking too much from the landscape, a term referred to as overgrazing, and he ordered the mass culling of thousands of elephants to address the problem. Regrettably, his follow-up observations confirmed that this was not the solution as the situation only worsened after the mass slaughter.

Allan began to realise that it was not the numbers of animals that were the problem, it was the way that these often managed groups were grazing that needed addressing. He talks about the role that large wild herds play in the landscape, grazing the land in a close group and always on the move, not returning to that area for maybe months or years and so allowing the vegetation to rest, recover and regrow as the soil is nourished from the trampled, ungrazed vegetation and piles of dung. His observations led him to realise that farmed animals needed to mimic these patterns if long-term damage was to be avoided, biodiversity maintained and soil fertility enhanced.

At Lynbreck, we would watch and observe the reaction from our land, changing and amending plans when needed, learning on the job of how to farm, with nature as our teacher. We were looking for more worms in our soil, more dung beetles breaking down manure, more species of grasses and wildflowers, more trees, more birds, insects and butterflies, all growing in greater abundance and diversity that would indicate our actions were continuously regenerating the land.

And while our approach was maybe different to others that farmed around us, it didn't really seem to matter. Most of the locals were just happy to see people wanting to work the land. For many, as long as we were producing food, as far as we could see, they weren't too bothered as to how we went about it.

We started to feel increasingly confident and clear about how we wanted our new lives and farming to look, but with growing lists of Lynbreck work desperately needing our attention, the challenge of time and money reared its head once again. By late summer of 2017 we were starting to reach burnout, trying to balance the growing demands of setting up our new business with the need to keep earning an income through our external jobs, often working sixteen-hour days just to try and fit everything in. Our new pigs had just arrived, we had fencing and building contractors to manage, multiple croft business admin jobs were piling up and, at least four days a week, both of us would have to leave the croft to do our second jobs. And all the while the bills kept rolling in as farm insurance became a new and costly necessity, electricity bills went up and we now had to pay an accountant to keep us up to speed with the completely new (to us) and intimidating world of annual business tax forms and quarterly VAT returns.

My stress levels had rocketed, and I was becoming increasingly cranky and snappy. Meanwhile, Sandra was having to manage her long list of commitments on the croft with various self-employed contract works off the croft, running busily from one task to the next as her own usually calm and steady nature became frayed at the edges. With our financial situation still a challenge, we were quite honestly afraid to walk away completely from the securities a steady job provided, lacking the confidence and assurance that we were ready to go it alone and still living on a month-to-month basis. In some ways, the final leap into full-time croft work was less risky than the giant stride we had taken to get to Lynbreck as we had the land and, technically, we had the dream. But, in all honesty, the fear of not realising the dream – what we had worked so hard for and what we had imagined so clearly – was crippling us into a strange purgatory. Almost paralysed with fear, we were defensively holding on to the last bastions of what we knew to help stabilise our unfamiliar life.

In the end, this was a reality check, a time to accept that we had been working ourselves too hard, pushing the limits of our physical, mental and emotional abilities, and now we were facing the consequences. It became difficult to differentiate what was important to us and what we felt we should be doing and that, at times, made even little things seem quite difficult to manage. Having time to cook meals, weed the kitchen garden, have a walk together in the woods, enjoy a cup of tea in the sun, host visiting friends and family. The irony was that all these little things were the reason why we had made such a big life move in the first place. The farming side came later but it was these core life choices and decisions that had brought us to Lynbreck and what, it felt, we were now sacrificing.

Sandra and I are both very instinctive people, knowing intuitively when something is right or wrong or needs changing. Our gut instincts agreed that the next big change on our journey would have to happen soon. When we've been in situations like this in the past, we find ourselves asking each other: 'What's scarier? Maintaining the status quo or making a change? Starting afresh or altering plans, even without knowing what might come next?'

On reflection, this was all part of the bigger transition as we gradually shifted from one way of life to another. I decided to reduce my work to two days a week and Sandra began the process of phasing out her self-employed contracts. The finances would of course be tight, but that was nothing new. While this was yet another risk, we gained the time and freedom to invest in our farming business and to recognise all the opportunities that came with that. We had a roof over our heads, food in the freezer, water in the well, firewood in the shed and a kitchen garden that was producing more and more each year, elements that we started to accept were what gave us tangible day-to-day security. Although the decision made us feel excited, we also felt conscious that the pressure was really on to make this life work, one where there was no manual with instructions to refer to – we would simply have to write it ourselves.

CHAPTER 9

Scratching Moss and Chasing Flies

It was a sunny Saturday morning in May as we drove up to a small town west of Inverness called Beauly. We were told to be at a car park on the outskirts of town for 11am sharp and to bring at least one large cardboard box. We had ordered our first three hens and it was here that we were to meet the chap and collect them. Upon arrival, there was a queue of people, all with cardboard boxes. At the front of the line was a van with a trailer, stacked with crates full of hens, and a man in white overalls who was busily opening them up, scooping out hens and bundling them into the boxes, the dark enclosure settling them into quiet.

This was Donald McDonald, a hen breeder from the Isle of Skye who had been recommended to us by some friends. Year on year he would raise thousands of chicks and every few months he made a handful of trips to locations around the mainland and isles to deliver 'point of lay pullets', or hens that would shortly be ready to lay their first egg, for their new owners. We stood in the queue, clutching our cardboard box and, before long we found ourselves at the front, giving him our name and order details and putting the box on the ground. With impressive efficiency he extracted three Rhode Rocks, a hardy Scottish breed with a beautiful black plumage that shimmers shades of petrol green in the sunlight, and with a bright golden orange blaze across their chest and around their neck. With relative ease, they went straight into the box and we quickly moved along, back to the car, delighted and excited with the first of our small, but no less significant Lynbreck livestock team.

The word farming can refer to multiple ways of producing food from the land and in our Highland landscape most farms carry sheep with a few cattle, the choice dictated primarily by the rugged upland terrain. The little experience that we had amassed in our previous lives came from a range of diverse environments and different countries, from Sandra's time working with cattle on horseback in Canada to our jobs as rangers in England and Scotland, where livestock played a part in conservation grazing. While our experience to date had been mostly observational, it was helpful to reflect on all these past learnings, using the little knowledge we had gathered to feed into early conversations about what livestock might work for our context at Lynbreck, weighing up the multitude of pros and cons of each in our setting.

From the beginning, the welfare of our animals would always command the top spot and this influenced heavily the livestock we chose and the way in which we hoped to work with them. Our plan was to ensure the basis of this relationship rested on a firm ethical and respectful footing where their sole purpose was not simply to feed us and our customers. Instead they would become a part of our team where, collectively, we would all work to benefit the natural environment and where our role would be akin to that of a coxswain at the lead of a rowing team.

And so it was a genuine high point to bring home the first members of our new flock of layers, initially for our own supply of eggs as part of our journey towards greater self-sufficiency. To some it may have meant 'just three hens', but to us it was yet another realisation of our dream and it became addictive to stand and watch them explore their new world in and around the homestead. Hens are opportunistic eaters and, just like us, are truly omnivorous. It was highly entertaining to watch them casually forage on a wide range of plants and seeds, perform comical standing jumps for a beakful of low dangling rowan berries, gobble their way fiercely through any vegetable scraps from the kitchen garden or run after flies at full speed and with incredible precision, snapping at midair as they hunted their aerial prey. Occasionally we'd watch one be chased by the others, the leader having successfully hunted a field vole and the followers desperate for a piece of the fresh meat. And their favourite thing was to do the hen shimmy, an action where they wiggle their whole body, scratching one foot after the next on the ground, looking for surface worms and grubs. The many hidden nooks and crannies around the homestead provided daily adventures and opportunities for hunting, running around the croft like mini dinosaurs.

We soon learned never to trust them as they became braver and bolder around us. On one sunny day, we'd been enjoying a few slices of cantaloupe melon, a real treat of something we wouldn't usually have. I made the mistake of setting my plate down, unguarded. Within seconds, a hen had snatched it and sprinted away. If we hadn't managed to get a picture of that moment, of a hen running off in the distance with a giant slice of melon clamped between her beak, no one ever would have believed it. We don't even know what happened to the skin, assuming the mini dinosaurs had probably gobbled the lot.

Just a few weeks after arriving, the new hens had each started to lay an egg a day, which more than supplied our needs. After adding some more hens to the flock that summer, Sandra built a shoebox-sized honesty box from some bits of scrap wood and she installed it at the top of our track, another small but very significant moment as the first of our produce was now offered for sale. We were only managing a few boxes a week, but we could see that the demand for farm-fresh free-range eggs from the passing traffic was strong.

At the start, all of our hens had names, beginning on a floral theme with Rose, Bay and Willow, followed by Herb, Meadow, Sweet, Sage, Thyme, Dandi, Lion, Blue and Bell. As more hens arrived, the names became more inventive. There was Wee Ghosty, a little White Leghorn that laid white eggs, who had the most impressive red wattle, bright yellow legs, and a bunch of tail feathers that sat horizontally and acted like a rudder in high winds, often steering her off course from the intended direction of travel. Then came the Supremes, three Cream Legbars that laid blue eggs, one of which we ended up calling Tina Turner as the similarity in hairstyle was uncanny, and she certainly had the sass. We had cockerels called Lion and Mango, and later on Basil, who strutted around like peacocks with all their ladies in waiting. As well as white and blue egg layers, we had a crossbreed known as an olive egger that would lay green eggs, in addition to more commercial breeds such as ISA browns that would lay brown eggs. When brought together in a half-dozen box, our eggs provided a rainbow of colour for our customers, the kind of eggs you can't pick up at the supermarket, not just in flavour but in appearance as well.

As we got to know their individual characters and quirks, there was one hen in particular who made an impression on us. It was one of our Marans, who initially was given the name

Sage, but which we soon changed to Football on account of the fact that her magnificent plumage and voracious appetite gave her a large, rounded body which waddled from side to side as she ran for treats. Football became quite an affectionate hen, sleeping next to or on us if we sat outside on a sunny day, and soon Sandra could pick her up and gently stroke her back, and her eyes would slowly close as she dozed off. Hens can be quite skittish but Football taught us that some could literally crave the attention of a good human keeper, showing incredible trust as she dropped all her vulnerabilities and replaced her distinctive chatter with a soft, sleepy snore. And, as her trust of us grew, we could scoop her up and allow any visitors to also have a stroke, giving many their first close encounter with a hen, an animal that provides such a popular food item but which many people would rarely come into contact with.

Within just a year or so we had built up quite a flock but, despite their best efforts, they simply could not keep up with the demand for eggs from our honesty box as word began to spread of the tasty, creamy, deep orange yolks from our free-ranging team. As well as enjoying their company, it was an incredible feeling to be able to eat our own produce. These little white and gold protein bombs were providing us with daily nourishment and though it was just a little step, it felt as though we had made another empowering leap on our journey towards self-sufficiency. However, our newfound optimism and confidence were soon to be tested.

'I can do this. I can totally do this. I just need to make it quick'.

Words I uttered to Sandra as I walked towards the side of the old croft house, my pace quickening as I cradled a small fatally sick hen in my arms. Sandra had noticed she hadn't been well for some time and, in spite of the best care she could give her, the hen was starting to decline and was in visible discomfort.

Hens have a phenomenal spirit and fight for life. Even when they are at death's door, they will do everything in their power to get up and out with the rest of the flock, not wanting to show any sign of weakness or illness that could make them more vulnerable to predators, a natural instinct that, despite centuries of domestication, manifests itself in times of need. The expense of veterinary intervention was one we couldn't justify and the only option was to put the hen out of her misery ourselves. This was a really big moment for us that acknowledged the shift we had made into keeping animals as part of a farming business. In our previous lives, we had only ever kept animals as pets. I grew up with family cats Snowy and Paws, and Sandra had a number of guinea pigs throughout her childhood. While in some ways we had become attached to our hens, this was a very different relationship and one in which new lines were drawn as it was time for us to make that first decision over life and death.

In preparation, I had spent hours researching how to home kill a hen humanely, my stomach still somersaulting at the thought. It wasn't the idea of killing her that was stressful, it was more actually doing it quickly and well, with my own bare hands. This thought alone made me nervous and I was worried that my self doubt would mess things up when it came to the time. But this was not about me. This was about ending her suffering and I just had to put my feelings aside and get on with the job. Still, I couldn't do it without wearing a pair of gloves.

After a few seconds, it was over. I picked up the body and walked across to the old cherry tree where Sandra had dug a little hole in which to bury her, popping a fragrant honeysuckle flower on the body before covering it with soil. We liked the idea that her body would now break down and feed the roots of this old tree, a fitting end that gave us some comfort. And, as we walked away, I turned to Sandra and said, 'Well, they say in farming where there's livestock, there's deadstock,' a lesson we would learn again and again.

As the flock grew, there were more hens that would have to be dispatched as occasionally some form of illness would set in and another decision would have to be made. Sandra became well read and experienced in monitoring their health, using natural remedies such as adding apple cider vinegar and garlic to their water a few days every month as an immune booster, and caring for those that were showing signs of sickness by providing simple treatments, which in many cases brought them back to good health.

And of course more hens would mean the need for more henhouses. During our visit to see Roger and Gilly on Lismore, we had been incredibly impressed with their small mobile henhouses on wheels that they moved through the fields into paddocks where their cattle had previously grazed. It was a set up that an increasing number of regenerative farmers were experimenting with on various scales and the logic they shared for doing so was very convincing. In the wild, flocks of birds naturally follow large herbivores as they move through the landscape, knowing that the piles of dung left behind attract all sorts of protein-rich grubs and insects. The birds scatter the manure as they hunt for these grubs and insects, fertilising the ground further with their own manure, before moving on. By replicating this relationship in a farmed situation there can be benefits for the hens, the cattle and the soil.

The hens get to feast on fresh grubs, learning quickly about the nutritious treasures a cowpat can hold. By eating fly larvae, which are often found in the dung, they can reduce the number of flies that will eventually hatch and can cause irritation to the cattle. Through the process of scattering, they help to break the manure down quicker, which feeds the soil beneath, as does the depositing of their own manure, which is rich in nitrogen. An additional benefit can come from the things hens love to do most, scratching the ground with their feet as they hunt for more snacks in amongst the vegetation. This was a habit we anticipated would be ideal to tackle the extensive patches of moss in our fields where diversity was very low and the land would benefit from animal impact. The idea of mimicking this symbiotic relationship and utilising their natural instincts appealed to us and, while we were still some way off to our first cattle arriving, we decided that a mobile henhouse would be the way to go.

A quick internet search yielded a second-hand caravan chassis for sale that would provide the ideal base for a small mobile henhouse. Having never built anything like this in the past, Sandra spent many following nights planning, designing, re-planning, drawing, re-planning and costing materials out until, finally, she had enough information and materials to bite the bullet and build the Eggmobile. After a week of long work days, head scratching, nail banging and the odd slip of a swear word, the paper drawings became a physical structure, built by Sandra with the attention to detail and meticulous planning that shines through in her natural carpentry skills. The Eggmobile was made entirely of wood to give it the weight we thought it would need to sustain the brutal winds that could hit us even in peak summertime and it was quite a moment when finally, on a hot summer July day, we hooked it up to the quad bike and towed it out into the field. There was no grandeur to the maiden voyage of the Eggmobile, only palpable relief that the quad bike had the power to tow it as we tentatively navigated the mole hill bumps and thick grassy tussocks in our field.

Just a few days later, twenty-five more hens from Donald McDonald were plucked out of cardboard boxes and bundled into their new mobile home. We set out a paddock with one hundred metres of chicken netting around the Eggmobile that we could electrify, which would function as a predator deterrent as well as a way to target the area that we wanted the girls to work. As we watched them settle into their new residence, it was interesting to observe and learn how they developed relationships with one another in a different setting, building their own social structure, commonly known as the pecking order. Their early interactions can seem quite brutal as they literally chase and peck one another, often detaching a beakful of feathers or even drawing blood. However, it is a structure that seems to give them stability and, once a new hierarchy is created, the flock relaxes, as bonds are formed and every hen has its place.

In addition to buying in young hens, we wanted to learn what it was like to raise chicks by hatching fertilised eggs. Sandra noticed one day that Tina from the Supremes had decided to go broody when she found Tina in a nest box, her body spread as flat as a pancake and her head tilted slightly down in an intense almost trancelike state. It's a very natural condition that only some hens experience as they become obsessed with wanting to sit on eggs to hatch – a good broody hen can be worth her weight in gold.

We bought twelve fertilised eggs from a nearby small-scale breeder and carefully slid them underneath her. After three weeks, a stream of high-pitched chirping noises flooded out of the main homestead henhouse window as ten little fluffy chicks (two of the eggs didn't hatch) appeared from underneath a very wary and now rather scrawny Tina who, for twenty-one days, had sacrificed her own needs to fulfil those of the new lives she had committed herself to rear. Every day she would talk to them, teach and fiercely defend them from the other hens and any doting humans who happened to get a little too close, flaring up her feathers like a large lekking capercaillie as she lunged forward with a warning – and rather terrifying – squawk.

Within a month, the chicks had begun to mature as their fluff turned to feathers and their confidence grew. They started to venture further and further away from the henhouse, mixing more with the main flock, but still returning to the safety of Tina's wings at night until, one day, Tina decided to rejoin the flock, leaving her chicks to fend for themselves, clearly reaching the point where she felt she had done enough. While we were a little taken aback, we had to accept that she knew what she was doing. A human reaction would be to see it as harsh but, in reality, this was an entirely natural process where the hen decided that the chicks no longer needed her. It was a situation that made us reflect on how humans can have a very natural tendency to impose our own feelings and emotions upon animals in our care, be it pets, livestock or even wild animals such as birds and hedgehogs that we become attached to as they visit our garden feeders. We can often accuse nature of being cruel, but maybe it's just our interpretation of what is a very rational, pragmatic and efficient system, the design of which is based on self-regulation, achieving the most beneficial outcome across the board for all life.

Inevitably, with any hatchlings, there will be some boys as well as girls and it was only ever our intention to keep the girls as future laying stock while rehoming a few of the boys, keeping one or two ourselves as future breeding stock and to act as guardian cockerels for our free rangers, dispatching the rest to go into our own freezer. It was a few years since we had eaten chicken, so while the prospect of a roast dinner was very appealing, we weren't particularly looking forward to the process we would have to go through for it. It's one thing to end the life of a sickly hen, but it's another to dispatch a young, strong, healthy bird and we had to learn to approach this with the pragmatism of nature, culling out the surplus and embracing it as part of the circle of life. Just a few weeks later we had the most delicious dinner of roast chicken where the skin was crispy and the meat beneath juicy, tender and packed full of flavours that neither of us had ever experienced with standard shop-bought chicken.

As we tucked into that incredible feast we felt a very deep, primal and even comforting connection to the world we were living in. This is a relationship that just a few generations past many would have had, engendering a natural respect and gratefulness for the food, where it has come from and the work that went in to producing it, be it meat, fruit, vegetables or honey. However, in our modern world, where a large proportion of animals destined for the food chain are raised in factories and fruit and vegetables are shipped from all corners of the globe, it's no wonder that we have lost that link to our food, which to us was helping to build a much closer and more tangible connection to nature.

By the following year, both of our henhouses were full to the brim as each team fell into a natural role and rhythm on the croft. We were learning just how underrated hens can be as working farm animals that can produce a continual supply of nutrient-dense, high-demand food *and* build soil fertility faster than any other of our livestock teams would be able to, as well as simply being a joy to have around (just not when we were eating melon). With both houses combined, we were at full residential occupancy at just over seventy hens and our capacity to produce enough eggs was still much lower than the growing demand we were experiencing. But we had no inclination or scope to increase the numbers, already committed to plenty of additional work around the croft as our other animal teams had begun to grow.

CHAPTER 10

Breaking New Ground

In the summertime, the British farming community come together at agricultural shows that take place all across the country. In our first summer at Lynbreck, we ventured an hour north to the Black Isle Show, the largest of its type in the Highlands, where farmers and crofters from all corners of the north of Scotland and further afield would converge, many bringing their finest specimens of livestock to enter into the various 'best in show' competitions.

It was a bit of a grey, overcast, damp day when we arrived at the showground along with the many other hundreds of cars, proving just how big an event this is in the northern farming calendar. We left the car in a field that was already so muddy it was like the scenes from a wet Glastonbury Festival, where vehicle wheel ruts were inches deep and many hundreds of pairs of feet had squelched the green summer grass into a brown, sticky quagmire. Fortunately, our standard foot attire of welly boots, a staple item for any aspiring or existing farmer and crofter, was ideal for the occasion as we wandered around the showground in relative comfort, enjoying the buzz of the festival-like atmosphere, saying many 'hellos' to the new friends we had made over the previous months.

At the end of one lane of tents, there was a little white marquee with the picture of a rather friendly looking ginger and black pig and a sign outside saying 'Oxford Sandy and Black pigs'. We decided to investigate further and headed for the entrance, giving ourselves a welcome respite from the mizzly rain and burgeoning crowds. A lady wearing an orange T-shirt with black writing and a black cap with orange writing bounced up to us, clearly co-ordinating the colour of her uniform with the colour of the pigs, a move we both found incredibly endearing. She introduced herself as Jane and also her husband, Adam, owners of Lyne Mhor Croft located just south of Inverness where they specialised in raising this native rare breed that had been deemed at risk of complete extinction only decades ago.

Jane took us around the pens at the back where there was a sow with piglets, all of which were looking very content and relatively unfazed by the events happening around them as they snuffled in their cosy bed of straw. She told us in detail about the breed and about the work they were doing as part of a dedicated group of UK breeders promoting the unique qualities of the Oxford Sandy and Black pig to ensure their survival. Most noticeably to us, she talked with such compassion and authenticity, showing real respect for the animals in her care that, on that day and in a very short space of time, we became rather fond of Jane and Adam, as well as their pigs.

That chance meeting turned out to be of pivotal importance as we were in the midst of planning a team of pigs for Lynbreck. We made a few visits to Lyne Mhor Croft, a small, vibrant crofting unit with sheep, goats, pigs, and a muckle of hens and ducks that, like our free rangers, seemed to be everywhere. Our goal was to learn and understand more about the practicalities and realities of keeping pigs, and the time that Jane and Adam gave us in those early days meant a great deal, leading to a new friendship as we saw for ourselves very clearly the respect they had for the animals in their care.

Within just a year of our chance meeting, we put in our first order to Jane and Adam for three weaners, young pigs at eight weeks old that have been weaned from their mum and are ready for the adventures of life ahead. There was a strip of land in our far field that was due to be planted the following winter with 320 native trees, which, when mature, would connect two existing woodlands and provide some much-needed shelter from the prevailing southerly winds. The area had been fenced off during the previous summer and ahead of us lay the job of preparing the ground for planting where dense grass, bracken and docks sat above a thick matt of roots that would provide tough competition for the little trees in those early days.

A number of years previously, I had the opportunity of visiting the Knepp Estate, a 3,500-acre landholding to the south of London, where the owner had taken the decision to move away from modern-day agriculture and to rewild the estate, retaining a number of native breed domesticated animals to fill the roles of their wild cousins. This idea of using our animals to fill the niche of a wild animal appealed to us strongly as it would allow them to express their natural behaviours whereas our role would be to target their impact in the areas that needed it. Pigs are great at disturbing ground, a job that was on our to-do list for the area we were preparing to plant. Rather than more days of arduous, backbreaking work for us, we decided a team of Oxford Sandy and Black pigs would do a much better job.

Our plan was to work the pigs through the area in small sections, encouraging them to target each paddock evenly and moving them weekly to avoid the ground becoming too broken up or the pigs becoming bored. Sandra set about designing and building a bespoke pig hut that could be taken apart quickly into sections and reassembled with relative ease, giving us lots of flexibility to move them without the use of a quad bike. Once the hut was ready, in place and filled with fresh straw, we used electric fencing to mark out the first area the pigs would work.

While in our minds they were no more or less important than our growing team of hens, it felt like another really big moment in our transition to our new farming life. There was something about bringing pigs on that had elevated us in our minds to a new level. Even just their sheer size, as bigger animals, and that they would provide us with meat and a new product to sell, felt like two substantial rungs up the ladder we were climbing. Although we were quite content to follow our own path and create our own route into farming, deep down we did look for some level of acceptance from the agricultural community around us and having a few more animals helped us to engage more in conversation with our peers. We could start to connect with them on things like the challenge of feeding rounds in the snow or which electric fencing to use – just little things but which to us felt very meaningful.

On the day we were to collect our new piglets, we felt quite nervous and endlessly grateful to Jane and Adam, who helped us load them ready for the journey back to Lynbreck. We'd never worked with pigs before, feeling very conscious that they had a reputation for being smart with a natural flair for breaking out of even the most robust of enclosures. But once unloaded and as they became familiar with their new house and place of work, we started to learn that if kept busy and entertained, they had no real desire to escape and we soon fell into our new routine.

It wasn't long before the winter came, the snow arrived and life was like living in a shaken snow globe where our white egg honesty box at the top of our track became as iconic to us as the famous Narnia lamppost that marked the entry point to another world. The days were cold and the nights fell well below freezing, and we became ever conscious from the warmth of our cabin for the pigs' well-being in this sometimes brutal climate. They were fed on a strict ration of two daily meals that were weighed out to the gram, but to which we had started to add an extra handful or two, justifying it in our minds that it would give them an extra bit of fuel to keep warm – and they certainly didn't complain. This decision, we found out later, was a mistake, and we had underestimated the sheer hardiness of the breed as they came back from the butcher with a thick layer of excess fat.

During their five-month stay, we closely monitored our pigs, assessing their impact day by day as we began to learn from our experience. As new pig keepers, it did come as a bit of a surprise to learn that, just like our hens, pigs are incredibly sociable beings, not just with each other but us humans too. Pigs will naturally explore with their mouths, inevitably meaning that wellies and legs can become interesting objects to nibble and nip, with the odd rogue bite bringing up some quite impressive bruising on our pale legs. But we soon found that their love of scratches became an ideal tool to train them, using these as positive rewards for not biting alongside several other cues to discourage them. This made being amongst them, even once they'd grown to a considerable size, an enjoyable experience as no longer did they run up to us just for food – the lure of a pig-tipping belly rub was also too tempting to miss out on.

By the end of their stay they had worked the whole strip, creating the perfect amount of disturbance so we could plant our trees, and it was time to prepare for our first run to the abattoir, a prospect that we were not particularly looking forward to. On the day before they were due to leave, we moved the pigs into a smaller pen with the livestock trailer at one end, which was full of straw, training them to go on and off by using food as an incentive. That night, the trailer replaced their hut, leaving the door open to give them the freedom to move on and off and helping to develop positive associations with the whole experience. Some folk might think us daft but we also each had a quiet word with them in our own ways, thanking them for the work that they did and the nourishment that they would give us. This is something we each do with every animal as they leave us, personally believing that on some deeper level of consciousness they understand. The next morning, we approached slowly, waking them up only for a quick morning snack and then closing the doors after the boys were quietly loaded and ready to depart.

We are very fortunate to have two abattoirs within fifty miles of the croft, a relative luxury in comparison to a lot of farm and croft set-ups. It is a worrying trend that these facilities are closing at pace around the country, putting many producers out of business as they simply refuse to put their animals through hours and hours of confined and what can be stressful road transport. In the UK, it is illegal to sell meat for human consumption that has not been killed via a registered abattoir facility and therefore the absence of these services is quite literally making or breaking small-scale farms like ours.

Arriving at the abattoir for the first time was quite a stressful experience, not knowing what to expect and unaware of the exact procedures. We joined a short queue of other vehicles and trailers until eventually it was our turn to enter the lairage, reversing the trailer back to a stone ramp that led into the building. The pigs walked off the trailer, following Sandra inside as she took them as far as we were permitted to do so before she turned back and we silently closed up the trailer and drove home.

That evening we experienced feelings of guilt and betrayal, starting to question everything: should we be eating meat?

Should we just home kill for ourselves and thus bypass the regulations that stipulate how your animals are to die? Deep down, we knew that in spite of good intentions, there isn't a single food choice made by any living organism that doesn't impact on some other form of life. We have never believed that to kill another life for our own personal sustenance is wrong, whether plant or animal. We know that animals are sentient beings with feelings and emotions but there is evidence that plants, although harder for us to relate to, have similar senses. In his book *Nourishment*, Fred Provenza, a professor emeritus of behavioural ecology, shares evidence that claims plants *see* wavelengths of light, *breathe* through their stomata and *feel* through their roots. He even talks of plants *smelling* and *tasting*, able to *sense* what is in their surroundings and *learn*, storing lessons in their *memory*. They are able to *communicate* and *build relationships* with other plants, alerting one another to predators by emitting warning chemicals through the air. Plants are so different to us that we often don't consider them to have traits that can be compared to that which we too experience, sometimes every day.

It made us reflect that the circle of life revolves around birth and death and, as humans, we have become very disconnected from the naturalness of that cycle, in particular when it comes to our food, as very few of us grow, raise or hunt – the actions that can build the deep relationships to what we then eventually eat. Reflecting on our experience, our struggle had not been about dealing with death, something neither of us fear, accepting it as a natural and necessary inevitability we all face. It was carrying the heavy weight of responsibility that rested on our shoulders to ensure that the animals we raised had a worthy life and a swift, respectful death.

We had arranged for a local butcher to collect and process the carcasses from the abattoir into various packages of meat – steaks, chops, sausages, burgers, roasts, diced, all of which we sold to individuals and families within our local community, keeping a few bits back for ourselves. And, with plenty more work to do, it wasn't long until the next group of weaners from Jane and Adam were ready to arrive. We planned to use the learning from our first experience to experiment with keeping pigs in a range of different set-ups and situations, exploring how to work with them to regenerate our land.

Our next team worked in field paddocks, disturbing and rootling though patches of moss and long grass. With every new move, it was fascinating to watch their preference for grazing the fresh greens before engaging their snouts with the ground, moving them only once they had worked the area enough, a judgement call that sometimes we made too early or too late as we learned by mistakes and successes. Afterwards, we replaced any turfs that had been flipped and scattered some native grass and wildflower seed on top of the fresh brown tilth to encourage colonisation of more diversity.

The idea in principle was good and, after a couple of seasons, there was a notable increase in floral diversity in the rested paddocks. However, what followed also was a spike in population of spear thistles, a pioneer plant that colonises quickly, throwing out its prickly arms to all sides, and that has an exceptionally deep tap root. It is crowned in the summertime with a beautiful bright pink flowerhead, attracting any number of insects and, as we found out, it is an absolute favourite for pigs to eat. We knew from the start that exposing the soil would result in nature trying to cover it up again as quickly as possible and, while there was greater diversity coming through in the rested paddocks where the pigs had worked, the boom in thistles had not been part of the plan. Yet again it was nature showing us who was in charge and the sometimes unexpected consequences of our actions.

With many more questions generated than answers provided, we decided to give the next batch of pigs a much larger paddock in our lower field, a landscape dominated by thick purple moor grass, rushes, bog myrtle and willow and quite typical of many parts of the more peat-based landscape of Scotland. This time the plan was to learn if their impact could increase the diversity and vigour of growth. We watched with interest as the pigs foraged on grasses and rushes, scratched against the small trees, wallowed in wet peaty hollows and created their own highways through the jungle of pig-height bog myrtle, a maze they could navigate with impressive speed and agility upon hearing the call for breakfast and dinner.

After removing the pigs in late summer, there was a marked flush of new growth the following spring. It made us question why more ground in Scotland is not used to keep domestic pigs in this way, instead of keeping so many confined in buildings when, with careful planning, monitoring and infrastructure, there is the potential to keep so many more in outdoor situations where they could have positive impacts. Many warn that pigs are escape artists, impossible to contain, but time and again our experiences were teaching us that if pigs are kept stimulated and busy in an outdoor setting with lots of space, a warm, dry place to sleep and ample food and water, they are quite happy to stay within the confines of their artificially demarcated territory, content that what they have is enough.

With summer experiments giving all sorts of food for thought, it was soon time to see what winter would bring as teams of pigs were moved into our woodlands, their mission once again to disturb the thick grass that dominated, opening up patches for other flora to colonise, including tree seedlings. The woodland work yielded results very quickly with little rowan tree seedlings starting to pop up in the birch-dominated woodland, a strong indication that our intended outcomes were beginning to materialise.

The Oxford Sandy and Black breed has been traditionally referred to as the woodland pig, and watching them in this habitat it was clear just how apt that description is. The pigs are in their element, their senses stimulated, shunning well-meant gifts of surplus vegetables from our kitchen garden as their nutritional needs appeared to be met by woodland foraging alongside their bought-in certified organic feed. And living in this setting brings out a delightful playful element with pigs running around the woodlands with sticks in their mouths, a pile of which would mostly end up in their house, an interesting collection that we would find as we refreshed their straw bedding.

One sunny, crisp day in the depths of midwinter, we decided to check on the pigs and spend some time socialising them. While sitting quietly in the woodland, we heard an almighty rush of sound as a huge flock of birds descended to the woodland floor. There were robins, wrens, great tits and chaffinches in such volume that the ground looked as if it was trembling in the dappled shade of the trees. The birds had converged on a patch of cleared vegetation that the pigs had just created, feasting on the woodland bounty exposed by the rooting snouts. As we sat and observed, we couldn't help but wonder if in the wild this would be a way of nature providing for the wintering birds, the pigs creating a virtual bird table that was stocked full of the freshest, protein-rich treats. It's said that one of the reasons why gardeners are so often followed by robins is because the little birds associate them with wild boar that rootle through the earth, exposing the soil, which has any number of subterranean worms and grubs for the bird.

Once an area had been 'pigged' and the animals removed, the plan was to give that land a much-needed period of rest and recovery. The impact of a pig can be very destructive if the ground is not allowed a chance to heal and we expected at least a good few years to pass before considering working pigs there again, using the recovery of the land as a guide to when might be appropriate. With that in mind, the decision was made to limit our annual pig numbers, carrying two groups through the winter and one group through the summer. A rough calculation helped us estimate that we could rotate this number around enough to ensure long rest periods, our priority sitting firmly with the long-term health of the land.

Looking back, our learning has also taught us that all pigs are true individuals with unique personalities. No two groups have ever been the same, with different dynamics, preferences and reactions to various habitats and even us. Just as we have our hen characters, we have our pig characters, many of whom have been given random names along the way. It's never something we do intentionally, especially with animals that are destined for meat, but they do sometimes end up with descriptive names as we can't help but become a little attached to individual characters.

In our second season of keeping pigs, Jane and Adam were not able to supply us with the number of pigs we were after, but their friend Bob, another local breeder, could. So in that group we had one called Bob's Pig. In other groups we've had Thera-pig, a pig so chilled out that sitting scratching him in the woods where he lived and worked was like therapy, in contrast to Luna-pig, a massive young boar who was a bit of a lunatic and once got so overexcited at feeding time that he nipped my leg so hard that it has left a scar to this day. And one year we had Micro-pig, a boar who had clearly been the runt of the litter and ended up around twenty kilogrammes lighter than the others he was with. We've had the 'Boys on the Bog', the

'Crazy Girls in the Woods' and the 'Boys of Bog Myrtle Jungle', all of whom have had their jobs to do.

But we also learned that pigs alone would not help to regenerate the land sufficiently. We were ready to go bigger again, to fill another ecological niche in our wilder farming model. That could only mean one thing – it was time for the large herbivores to return to Lynbreck.

CHAPTER 11

Hardy Beasts

'Come on,' shouted the farmer, calling the cattle in for their afternoon treat. All of a sudden, the group in the distance looked up, their ears pricked at the noise they heard every afternoon, one which they knew, without hesitation, the meaning of. The quiet, sunny afternoon was soon shattered with the rumbling of multiple hairy, mostly horned beasts running down the hillside, their red hair blowing in the wind and their hooves kicking up small plumes of dust, their enthusiasm for the call palpably clear.

Upon reaching the feeders, they tucked into the fresh grains that the farmer had poured into the troughs, their lips and hairy muzzles soon coated in seeds as they ate as quickly as they could in case their neighbour might shunt them out of the way. All except one. A little black, timid, almost frightened-looking heifer, stood at the back. Even from a distance we could see her black eyes were wide open, her head twitching as she stood on guard, ready to run away at any moment. After a few minutes she tentatively made her way towards the feeders, before jumping back again, one look from one of the others enough to shoo her away.

This terrified, cowering, near scraggily wee beast was Ronnie.

In those first few months when we were planning our hill-planting project, we had a visit from a lady from the RSPB who wanted to find out more about what we were up to. During a wander, we mentioned that our hope was to get some cattle and she recommended getting in touch with a chap called Bill, a local farmer who had been doing great things for nature. I dropped Bill an email with my phone number and, within about thirty minutes of pressing send, the phone rang.

Bill and I must have talked for nearly an hour that day. When he got the email, he noticed that my surname was the same as his, a relative rarity in this part of the world. It turned out his family originated from a township not far from where my mum and dad came from, and the chances are that at some point in history we were actually related. A few weeks later we arranged a visit to meet Bill and his wife, Steph, popping over for a short visit, which turned into a five-hour bonding session. From that point on, we referred to one another as family.

Bill and Steph were just retiring from farming but had a small fold of Highland cattle that they were preparing to sell on to an incoming tenant. They took us to see them, letting Sandra join Steph one day as she was giving them their afternoon feed and talking us through all the different equipment needed. It was really helpful having them spend the time with us, offering positive and encouraging thoughts and advice as we tried not to become too overwhelmed with all the new information.

Our meeting couldn't have come at a better time as we were trying to decide on which breed of cattle would be best for our land at Lynbreck. Once again we found ourselves in a whole new world, realising that we actually only knew a handful of breeds. We definitely wanted to go for a native breed but the more we learned, the more our options grew and grew.

There were Highland cattle, Shorthorns, Luings (a highlander crossed with a Shorthorn), Galloways, Belted Galloways, Shetlands, Dexters, White Parks, Longhorns, any of which could have worked and some of which came with horns while others didn't. We had always been drawn to Highland cattle but became wary when other farmers would say things like 'they're wild animals' or 'they'll rip all your fences out', giving us sleepless nights as we imagined these hairy beasts causing havoc with our recently installed grant-funded fences. But the ones we met at Bill and Steph's were not wild and all of their fences seemed perfectly intact. They taught us that these cattle were placid and relaxed and thrived on regular human contact, which helped keep them calm and quiet to be around.

But then, there was Ronnie. Bill told us that her mum, also called Ronnie (it's a tradition for Highland cattle females to take their mum's name), used to think she was the big boss until one day the other members of the fold put her in her place and she fell to bottom rank. There must have been something in that which was passed on to young Ronnie as she shirked to the back, keeping herself to herself. But she did intrigue us with her dark, thick black coat, small white horns with black tips and a white patch on her underbelly.

About a year after our first meeting with Bill and Steph, we popped round for lunch and a catch up. They insisted on showing us something in the barn just as we were leaving. Ronnie was inside with two other red Highlanders, each named Flora and of the same age. Bill turned to us with genuine emotion in his eyes and said, 'We would like to give you Ronnie as a gift from us both.' We were stunned to say the least.

On the one hand, we were overwhelmed at their kindness. Two people who, just a year ago, had been strangers, were now showing us incredible warmth and generosity – that meant much more to us than we could put into words. On the other hand, we were terrified. Ronnie was going to take some work, her edgy aura making us feel slightly nervous and uncomfortable as we knew she would only get bigger and her horns more pronounced. And we had at best a tiny amount of experience of working with cattle. But, we never hesitated in accepting the kindness of our new family, talking pragmatically about Ronnie in farming speak: 'If she doesn't fit in, she'll just have to go.' Our commitment to Highland cattle for Lynbreck was sealed as we decided to buy the two young Floras as well as a steer who would be the first of our beef cattle. He was a calm and docile beast whom we ended up calling Mr Steer, a name that would bring us no prizes for inventiveness or ingenuity, but which just worked. To build up the numbers a little more, we bought another two young steers from a neighbouring organic farm who came with magnificent sets of horns and became known to us collectively as Junior and Junior. With a total of six in all, our new fold of Highland cattle was complete.

Highland cattle are incredibly hardy beasts that grow a thick shaggy coat in the winter to keep them warm in the harsh conditions, which they shed out for the summer using trees and exposed rocks as their hairbrush to scratch and comb out the old hair. They eat a wide range of vegetation, from grasses and wildflowers to shrubs and tree leaves, and are naturally built for all weather conditions, enabling them to be outside 365 days of the year. Their beauty and iconic image can often do them a disservice, sometimes having the association of being 'hobby' cattle for 'hobby' farmers and demoted to being the face of toffee bars and advertising campaigns. However, to us, these animals were much more than just a pretty face, acknowledging their natural resilience and strength and seeing the perfection in their suitability to become a part of our team at Lynbreck.

In my previous job with the Borders Forest Trust, I had been fortunate to meet a gentleman called Roy Dennis, a virtual giant in the world of conservation and a pioneer of habitat restoration, mammal and bird reintroductions. Roy had previously lived and worked in Abernethy Forest and in 1998 published a paper entitled 'The Importance of Traditional Cattle for Woodland Biodiversity in the Scottish Highlands', which we came across in our early Lynbreck days. He wrote about the potential benefits of keeping cattle that could fill a niche in our natural ecology, the importance of dunging for soil health and insects, the creation of different habitats as a result of their presence and movements through the landscape, and the recycling of plant material through their grazing patterns and preferences. Roy made a compelling case for the use of hardy, native breed cattle in small numbers at low densities that would also ultimately provide exceptionally high-quality beef that could be marketed and sold at a premium.

We wanted to run our Highland cattle in as natural a grazing system as possible, where the plan would be to keep them moving regularly, utilising their instinctive behaviours. This would help to ensure the spread of a fairly even deposit of dung and urine across the fields and, with hardy cattle such as Highlanders, there would be no need for expensive housing in winter. We would mitigate against the ground becoming too broken up and mucky by carrying low animal numbers on a grassland where the plants had deep root systems, able to hold the weight of the lighter-framed Highland cattle. And the plan meant avoiding any unnecessary soil compaction, a condition caused by repeatedly carrying heavy weights like large groups of heavy animals or vehicles, leading to the oxygen literally being squeezed out of the soil and rendering the ground lifeless. In contrast, our aim was to have light, crumbly, aerated soil that, when the rains came, could literally soak up water like a sponge, water being an essential natural resource that we wanted to store in as many ways as possible for the plants above.

And the most important part would be ensuring that the area where the cattle had just grazed would be allowed sufficient time to rest and recover. When a plant is grazed by an animal, it needs time to fully regrow, and if it is not given that time and is grazed again, it draws on root reserve energy to keep going, a process called overgrazing. When this happens repeatedly, the plant will eventually become stunted, producing less biomass and it will ultimately die. Through careful planning, we could avoid that, helping to keep our pasture healthy and vigorous.

Our responsibility would be to oversee those movements, which, in the summertime would mostly be daily, factoring in considerations such as availability and diversity of forage, water and shelter, as well as seasonal things like calving and weaning. This consistent interaction would also give us the opportunity to regularly monitor their grazing impact, making changes to our plan when needed, building knowledge and experience as we went. Our now go-to animal author Fred Provenza claims that herbivores, if given the chance, will graze on up to fifty different plant species in a single day, based on their need for specific nutrients and minerals. Our pastures had the beginnings of a diverse flora forage platter for the cattle to pick through and, we hoped in time, their work would help to grow the diversity of what they could choose from.

With ambitious but carefully thought-out plans in place, it was soon time to test whether our strategy for the Highland cattle would indeed work, considering our inexperience with such large animals and wondering what the ever-nervy, unpredictable Ronnie would make of it. We weren't even confident that we would get her on the trailer when it was time to collect them but, to all of our surprise, Ronnie was the first to load, followed by one of the Floras, so we quickly shut the tail gate and began the first trip back to Lynbreck (we were only able to fit two at a time in our small livestock trailer).

This was us well and truly on our own now. Once these first two were unloaded they were our responsibility and the time for talking and planning was over. We drove slowly down the bumpy entrance track at Lynbreck and Sandra jumped out to open the gate of the new fenced training paddock next to the homestead. I reversed up and, once in place, we blocked any gaps at both sides, not trusting either of them not to bolt. Sandra pulled down the ramp and carefully opened the slatted metal gates as I stood back, not wanting to be a distraction. Without hesitation, Ronnie was the first to exit the trailer, her hoof becoming the first of any cattle that had graced Lynbreck for a number of years. To our relief, she wasn't panicking, freaking out or going crazy. She was alert but calm and Flora followed suit. We collected the others shortly after and finally the two Juniors a few days later, and, before long, realised that something incredible and unexpected had happened.

Ronnie had become the boss.

We've heard tales before of cattle arriving at a new place and the first one to step off the trailer claims the land as theirs, but this was not something we expected at all of the shy, anxious, nervy Ronnie. However there was no doubt in our minds that this is what had happened. When we brought them hay, Ronnie was the first at the pile, tipping the point of her horns towards the others as a warning that she would get first dibs. If there was a queue at the water trough, Ronnie would make her way to the front, the others parting like a Red Sea of Highland cattle as the recently crowned matriarch commanded liquid refreshments.

To get the desired outcome with our grazing, we had bought a load of new electric fencing kit to make a number of smaller twenty-four-hour paddocks within our field. As well as setting up the fencing, our role would be to work as cowherders, carefully monitoring the land and deciding how long it would be before the cattle could return to previous grazings, a skill we learned quickly was an art taught by experience rather than a science taught by books.

Sandra had been training the cattle during those first few weeks, setting up electric fencing lines in the paddock, which, after one or two zaps, they soon learned to avoid. Sandra would shout 'come on', a vocal instruction that they knew to follow from their previous farm, associating it with a reward of food, before feeding them hay. Bill had mentioned to us that there are two things that all animals, including humans in most cases, generally respond to: food and sex. The sex element was something that would require a fully functioning bull as our boys had been snipped. But when our girls came into season, Mr Steer and the two Juniors certainly gave it a go, their lack of male tackle doing nothing to deter their horny urges and natural instincts as they mounted the backs of the girls who were indifferent to say the least. For a couple of days every month we would watch them follow whichever heifer was in season like lovesick puppies and then, once this time had passed, the boys would return to their usual aloofness.

But the food bit we could definitely use as an incentive and it worked a treat. After just two weeks of training where we weaned them off grain on to just hay, they would follow Sandra to wherever she would call them in the paddock for their breakfast and evening meal. This helped to establish her as the true head of the fold, even above Ronnie, as she became the provider of food, water and shelter. She also started to learn about building a mutual relationship founded on respect, an important quality not least for the safety of a livestock handler. But even with the cattle now ready for grazing action, we still had a final hurdle to overcome.

'How big should the twenty-four-hour paddock be?'

'I'm not sure.'

'Well, how much do cows eat in a day?'

'I'm not sure.'

We stood there looking at the field, scratching our heads as we tried to imagine the size they would need. We wanted to give them a big enough area that they would have plenty of food to eat, space to lie down and lots of opportunities for cowpats. We felt a lot of pressure to get this right, eventually deciding to err on the size of bigger rather than smaller, thinking we had to start somewhere.

After just a few weeks, morning moves were done and dusted in just a handful of minutes as the cattle responded to our calls, instinctively ready to move. We learned to read the land, watching the impact the cattle had and using this and the amount of forage below to tweak the sizes of paddocks every day. We became electric fencing ninjas, able to set them out, move cattle, move water and take down the old fences in just an average of 30 minutes a day.

The time investment in our cattle was starting to pay dividends as the fold fell into their daily routine of eat, sleep, move, repeat. Sandra would use these daily interactions to visually check on their condition, allowing us to monitor their health so that if there was an issue we could pick up on it early and act where necessary. We began to get used to them and their quirks as they started to feel more comfortable around us, nurturing a respectful relationship that was based on calm and consistent interactions, the latter of which in particular is a very important and often undervalued aspect of animal handling. Animals are naturally consistent in their use of body language as that is one of their primary modes of communication. In contrast, humans tend to rely more on verbal communication, meaning that we often have a very poor awareness of our own body language, which to our animals can come across as quite erratic. By learning about an animal's behaviour and consciously applying our own body language in a consistent manner, we start to 'make sense' in the eyes of our animal counterparts, helping to garner a more trusting relationship. Our voices can play an important part as well, as either a cue or calming influence – however, our body language will always convey the truth.

And, of course, the fact that they had big pointy horns made low-stress handling even more important for us. All of our females had sets of magnificent iconic Highlander horns that swept out from the back of their heads and up, ending in a smooth rounded tip. In contrast, the boys were without. It's common practice for males to have their horns removed, primarily as a safety precaution for the farmer but also because many abattoir facilities will either impose restrictions on accepting horned cattle or, in some cases, will not accept them at all. That means running a beef herd with native, horned cattle can be very problematic, a restriction that really frustrated us. We knew that there was evidence to show that horns can be a part of a cow's natural biology, connected to bodily functions such as temperature regulation and digestion and an important method of communication in establishing hierarchies where subtle movements with their horns can settle their ranking. To remove them is to remove this part of their very nature and it's something that still really bothers us today.

With daily moves going more or less to plan, it was soon time to face the ultimate test of just how low stress and calm a relationship we had with our Highlander team. For the first time ever, we were about to bring them into our barn and through our new cattle-handling facilities, but having no idea at all how they would respond. A cattle-handling system is a series of gates and compartments made up of hurdles that eventually leads to a crush, a large metal crate with two bars that would clamp either side of their neck and hold them in place. Although it sounds and can look intimidating, it's an important device that provides safety to both handler and beast during inspections and treatments, and the theory goes that with consistent, low-stress handling the livestock soon get to know the routine and enter the crush without fear.

Sandra had been reading the works of Temple Grandin, an inspirational American animal behaviour expert and autism spokesperson, whose own experience of autism, as well as scientific research, had enabled her to design livestock handling systems that significantly reduced stress in animals. Her autism-related exposure to anxiety and heightened sensitivity to her surroundings had given her a deep insight into how animals and livestock in particular perceive their immediate environment. While our set up would be quite small, Sandra assembled it with her usual careful thought and precision, taking what she had learned from Temple and applying the same principles that would encourage the cattle through the handling system without visual distractions and create a natural flow.

On the day before we were due to take them through our new system, Sandra opened up the barn so that the cattle were able to go in and out at their own leisure to explore. These are animals that are never housed and are unaccustomed to being confined in an indoor space, so giving them the time to explore at their own pace helped to alleviate any fears through building familiarity. And for us, it was fascinating to watch. We hung back not wanting to be a distraction, standing on the decking in front of the cabin with our binoculars. Slowly, one by one, they would approach the entrance, their eyes wide with inquisitiveness, before meandering inside for a look. Sometimes they would be in there for a few minutes, leaving a giant cowpat as a calling card that would splatter as it made contact with the hard concrete floor. But before long, the barn became 'no big deal' as they settled nearby, chewing the cud.

The next day, we secured them in the fenced area at the front of the barn and began to move them quietly and slowly inside. Our hearts were beating so loudly it was like the drums of Lynbreck but fortunately only a sound that reverberated between our ears. There was no need for shouting or grand arm-waving gestures and we certainly had no plans to rush, conscious to avoid causing undue fear, a reaction which could cause a fight-or-flight reaction from them, putting us all in potential danger. We simply held a partially unravelled and unconnected length of electric fence wire between us and, as we walked towards the fold, they instinctively walked away from the line and were guided into the barn.

Once the gate was closed, Sandra encouraged them into the back of the barn where they started to flow into the race section of the handling system, a gradually narrowing passageway, until eventually the first steer stepped into the crush where I swiftly pulled the lever, clamping the bars around his neck to secure him in place. He had an ear tag that needed replacing, a common occurrence with a Highlander as they vigorously scratch their heads on the side of trees and fence posts, dislodging the yellow plastic disc, which holds their unique identification number and is required by law. Sandra had the new piercer and replacement tags ready and, for the first time ever, reached through the bars until she was touching the head of this hairy animal, all the while talking to him in a quiet, soothing tone, a voice he had always associated with positivity. I knew her heart rate would be sky high as she wanted to make the experience as quick and stress free as possible but, like with everything we tackled, the first time was always the hardest as textbooks can only prepare you to a certain point. I, on the other hand, couldn't even watch. I've always had a funny relationship with anything medical, harbouring a deep fear of needles of any kind, so instead focused my attention on looking ahead, maintaining concentration on the job in hand and ready to respond to any assistance Sandra would need. After a bit of fiddling around, Sandra wanting to get it just right, I heard the clunk click of the piercer securing the new tag and we all breathed a very loud sigh of relief.

Before releasing him, Sandra gave him a close-up visual once-over, checking his overall health and condition before rounding off the experience by holding up a bucket of lucerne nuts – little pellets of crushed dried lucerne, also known as alfalfa – which he tucked into with delight. This last act was probably the most important of all as the steer would now leave the crush with a positive final association, helping to reduce any fears or anxieties for the next time. As she moved the rest of the fold through the system, we repeated the experience with all of them until, finally, everyone was through. Sandra allowed them access to the inside of the barn for the rest of the day, should they decide to go back in for further investigation but entirely at their own choosing, which helped to round the whole experience off as 'no big deal'.

'Well, that went well,' I said as we walked back to the homestead.

'Yeah, I'm really pleased with that.'

And she had done a brilliant job. As much as it was a team effort, Sandra had worked so hard to get it right for these hardy beasts, her respect for them coming through in spades. She had spent hours preparing, so to have it all go to plan gave us the biggest high for the rest of the day.

A few months later we tried to remember that boost as we loaded Mr Steer, now destined for the abattoir, onto the trailer. Some people find it harder to take pigs to the abattoir, citing their obvious joie de vivre like an excitable puppy pulling on their conscience and heartstrings, but we found it just as difficult to take our cattle. We noticed how they were the most introverted of all of our animals, probably to do with the fact that they are ruminants and spend a lot of time digesting their food by chewing the cud, a natural internal biological process that has a lovely introspective quality about it. But that isn't to say that they aren't full of character and can show a lot of curiosity, often standing mesmerised as they watch any kind of activity going on in their surroundings.

Throughout our carefully planned summer grazing, we had been stretching out our available forage for as long as possible, utilising one of the many superpowers of Highland cattle whereby they are able to live on nutritionally poorer vegetation and still thrive. But, as autumn turned to winter and the last of the standing grass was grazed, we unwrapped our first hay bale, unleashing the warming smell of summer grasses and wildflowers and taking armfuls out to our fold who were starting to vocalise their discontent with the forage scraps left in the field. Ronnie, in particular, would look at us with immense discontent. She's always had a healthy appetite, her frame perhaps carrying a touch more weight than the others as she certainly enjoys her food, yet another attribute that just became part of Ronnie. They all tucked in with such delight as we stood back, watching them eat every last stalk before collapsing in heaps onto the hard, frosty ground, chewing the cud with utter contentment. It was from this point on, when we watched them lie down with full bellies that we affectionally referred to them as the 'lumps'.

We began to use our twice daily hay allocation to continue our regular cattle moves, feeding them in different locations, but this time in much larger paddocks so that they would always have access to shelter from trees should a winter storm come in. Because the cattle wouldn't be in any one place for a long time, this helped to avoid our ground getting muddy and compacted. Any scraps of hay that they didn't eat would simply break down, feeding the soil as dislodged seedheads were trodden into the ground, planting them to sprout new growth in spring.

We also started to experiment with a technique called bale grazing, a tactic of simply putting a hay bale in a spot in the field that we'd identified as needing a bit of a fertility boost, taking off the netting that holds the layers in place and letting the cattle eat from it directly. Once prepared, we would bring the cattle to the bale, a time of obvious excitement as the boys would head-butt it or the girls would dig their horns straight into the side, sometimes flipping the bale, before they settled down to feast. It would take our fold between two to three days to finish their 'all you can eat' buffet where the end result was a large circular matt of hay a few inches thick. We had to fight the urge to rake the leftover hay off, subduing doubts sowed by others that the 'waste' would 'smother' the grass in the summertime, knowing that we had to leave this 'wasted' hay with patches of dung to break down into the soil. The following summer, these patches were transformed into lush and diverse forage, buzzing with insect life as little field voles utilised the new cover to move around their territory. Through this very simple technique, we were slowly starting to increase the amount of diverse grazing for our Highlanders, now seeing 'hay waste' as soil food and a trickle investment into the prospering bank of soil.

MOOOOOOOO! MOOOOOOOO!

The sound was booming up from the flats.

It was my turn that evening to lock up the hens and I heard the sound as soon as I stepped out of the cabin at about 10.30pm. I knew the cattle were out on the bog and this 'moo' was different to the ones we had heard before like the 'I'm hungry' moo or the 'I'm excited and feeling frisky' moo. This was distress.

I ran back into the house and told Sandra. 'I think something's wrong with the cattle on the bog,' panic seeping out through every word. 'Oh god,' came the reply, and I knew why. Both Flora and Flora were pregnant after an eight-week 'holiday' with a handsome Highland bull from a nearby organic farm the summer before, and our thoughts immediately went to them.

'But that would make them over a month early. It can't be that,' I said breathlessly as we hurtled down the field. Just a few days earlier I remember Sandra commenting, 'Both Floras look as though they are starting to bag up,' a term used to describe a cow's udder filling out with milk and a sign that calving can be imminent. However, we'd been given an approximate date by the vet when they were scanned and there were still weeks to go. Our plan had been to bring them up near to the barn in a few days' time in preparation.

As we approached cautiously, one of the girls was standing, visibly and audibly distressed, and the other was lying a few metres away, her large core frame heaving back and forth, giving the impression that she was going into labour. It was clear she did not want us nearby, hoisting her heavily pregnant body onto her legs, and wandering off to the far corner of the bog. Highland cattle are notorious for wanting to calve in secret and they have a phenomenal ability to pause the birth if it's not quite the right time.

We managed to separate the rest of the fold into the adjoining field with some bags of hay to give the girls the space they might need, before returning to the flats, moving a little closer to the distressed heifer. 'She's given birth. She has a little calf.' The heifer was in fact now a new mother cow herself, standing above a beautiful bright red bundle of fur, nestled in a bed of heather and obviously alive.

The excitement and relief was huge but short-lived as our concern became focused on the other heifer that had moved away. We decided to leave her alone and give her the space she clearly wanted. By the time we headed out again it was nearly midnight as we scrambled along the heathery bank in near complete darkness, eventually stopping a good distance from where we thought she had relocated to.

Sandra shone the torch out across the flats, sweeping the shard of light slowly, eventually stopping when it met the bright eye of the heifer twinkling in the powerful beam. Again, conscious not to bother her by moving closer, we used the torch to search the surrounding area but could see nothing, eventually switching the light off and settling onto a ground mattress of springy heather under a clear, star-filled sky to wait.

Within a few minutes, we heard the most almighty 'whoosh' and we knew she had just calved. After waiting a couple more minutes, Sandra shone the torch out into the sweeping black, conscious that we didn't want to spook her away from the calf at a very crucial moment. When the light found her, our hearts sank as we could see the calf lying behind but neither of them had got to their feet.

'I'm calling George,' announced Sandra. George was our neighbour who we had on standby and who encouraged us to get a closer look, but with a firm warning to keep a safe distance because new mums can be extremely protective, especially Highlanders. We started to move slowly, the lanky stems of heather wrapping around our legs, causing us to tumble as we tried to navigate the undulating terrain beneath us in the black void. We climbed over the fence and Sandra edged closer and closer until eventually she was right next to mum and calf. The realisation of what was going on hit her hard.

The heifer had given birth very easily and, it being her first time, she obviously didn't understand what had just happened. The calf had slid out with the sac over its head and lay waiting for mum to get up and start licking to encourage its circulation and clearing its airways. But mum didn't move. And so, within just a few precious minutes of life, the newly born calf suffocated in the very membrane that it had grown in and had been protected in for months. Sandra immediately cleared the sac from its nose and mouth but she knew it was too late. By this stage, Flora's maternal instincts kicked in and she lifted herself off the ground and immediately started licking her newborn. Sandra backed slowly away, her heart sinking with every step as she watched the cow try to bond with her firstborn, knowing the outcome would be futile.

Our hearts were full of the heaviest sadness as we left Flora for the night and returned to the cabin, climbing into bed for what we knew would be a short, sleepless night. At first light, we returned to check on both new mums, our spirits raised as we saw one newborn up and suckling on a very proud and protective mother, before emotions tumbled as we found the other pacing around, bellowing in mourning as the understanding of what had happened started to sink in.

We decided to let the rest of the fold out to join them as they were clearly unsettled by the night's events, aware that they had a role to play in all of this. Ronnie led the charge, heading straight for the first cow and calf, inspecting the new member of the fold, while the cow stepped aside, in acknowledgement of the hierarchy. This was followed by all the steers investigating, a few taking more of an interest than others. Some of those bonds formed on that first day were clear to see in months to come with the younger steers stepping in as playmates for the only calf.

The fold then headed up to where Flora was with the dead calf, investigating the situation before settling in the area. Over the course of the next few days, the grieving cow visited the calf regularly, in between wandering off but still not joining the others at feeding times. We'll never know to what extent the fold was consciously engaging in a group mourning ritual, but whatever it was, we knew to stay back and let them do what came naturally to them.

At that time and even looking back now, it was one of the hardest experiences we have had since being at Lynbreck, replaying the night over and over in our minds, wondering if the outcome might have been different if we'd intervened sooner or would that in turn have had unintended negative consequences? We felt such a weight of responsibility for these lives that worked alongside us and we felt our own deep sense of sadness. It's one of the contrary elements of farming that's best summed up in that sentence once again: where you have livestock, you have deadstock. Animals dying, whether old or newborn are a part of the deal, and that's one of the things you sign up for and a fundamental part of what you just have to deal with. When they suffer, we suffer. When they are content, we are content. Their health is our primary concern.

It's an experience neither of us will ever forget but it was something we had to accept and move on from. Either you do that or you stop farming, simple as that, and, while Sandra is much better at handling these situations than I am, I definitely thought about walking away from it all. But, within a few days, Flora started to integrate slowly back into the fold as their lives returned to a new normality. We took huge comfort in watching her become a nanny for the new calf, looking after it as the mum would go off for some food, able to utilise some of her natural maternal instincts. And the calf was flourishing. Every day she got stronger and stronger, doing what we would call 'crazy runs', where she would just get a burst of energy and run around as the others watched with what looked like a sense of bemusement, clearly far too old and mature to partake in such energetic frivolities. While consciously trying to not impose our own human emotions onto the situation, we watched as they all just kept on living, back into the routine, but now with an added bit of bounce thrown into the mix.

Once again we were reminded that these animals, like all of our livestock at Lynbreck, are never just a cow, or just a pig or just a chicken, they are magnificent, living, breathing creatures that are strong and resilient and live each moment, second by second. They inspire us and make us smile every day, and for their sakes, and for ours, we knew that we had to just keep on going.

CHAPTER 12

You Can't Eat Trees

A few days before we were due to finish our jobs in the south of Scotland, I was having a final chat with a farmer whose land shared a boundary with Corehead, which was one of the sites that I worked on. Corehead had been an old hill farm that the charity Borders Forest Trust had purchased a few years before and had retained traditional sheep farming on one side and planted native trees on the other. Inevitably, our conversation turned to the land and she said, 'You can't eat trees,' a comment on the removal of farmed sheep from the land. I went home that night and told Sandra who said, 'Did you tell her that you can't breathe sheep?' We had a bit of a chuckle, but I decided not to go back and share it with her, thinking she might not appreciate our humour. While I understood what she was saying, I never actually thought she was completely wrong until having cattle taught us that, strictly speaking, she wasn't absolutely correct.

It was an early warm sunny morning in July as Sandra walked across the field to move the cows for that day. The Cairngorms were looking magnificent as beams of light from the rising sun hit the tops with shrinking patches of snow from the winter past still visible, hanging in the northern-facing corries. We were in the middle of a very hot, dry spell of weather, the kind of temperatures that keep even the midges at bay as we dined outside in the evenings, moving our table onto the small bit of decking in front of our cabin and feeling like we were in deep central Spain.

It was just a few months into our first cattle-grazing routine and, so far, things had gone without incident. On this particular day, they were due to move into a paddock in the middle of the field, something that normally we wouldn't have thought twice about, but with temperatures forecast to soar to 30°C, we had a few concerns. On previous hot days, the cattle would spend a lot of time in the shade of trees either on the edge of or within that day's paddock, but after today's move they wouldn't have that option.

At around midday when Sandra went back over to the cattle, they were standing up and panting, displaying classic symptoms of heat stress. Highland cattle do shed out their thick coats after winter but they are still pretty hairy in peak summer nonetheless. While the hair reduces from their body, they tend to grow these magnificent fringes, a biological miracle that helps to keep the flies from bothering their eyes. Without hesitation, she removed one of the electric fence lines and called them across to the woodland edge which was casting a cooling shade over the adjacent paddock, and topping up their water supply should they need it. Within just a few hours, the panting had stopped as their body temperatures started to readjust and they began to graze again as normal.

The advent of autumn brought with it the seasonal storms that would hit us with such force that they caused the walls of our cabin to shudder, leading to many a sleepless night worrying about roofs staying on buildings. The winds can blow so loudly up here that it's sometimes impossible to sleep, a mixture of noise and anxiety causing us to lie awake all night, turning every now and then to the other and asking 'are you awake?' as you feel almost frightened to be the only one experiencing it. And the cold can be brutal. On windy days we keep our outside jobs to a minimum, the gusts making it virtually impossible to stand, let alone work, but it's the vicious chill that really forces you inside. On those kinds of days we have the wood burner lit from early morning to try and get some heat into the cabin. With no central heating, it usually works a treat but when the winds swirl around us, the heat from the fire is sucked straight out of the chimney as we put more and more layers on to try and combat the increasing chill.

Those were the times when our thoughts turned to the animals. The hens and pigs have their cosy, straw-filled huts so they were generally not our worry, but our fold of Highland cattle were always outside. After a particularly brutal night, we headed out at first light and found the cattle in and amongst the trees, knowing instinctively where to take shelter from the high winds, snow or driving rain that would lash down with such ferocity it felt like being pelted with little darts. Our neighbour Hamish would refer to woodlands as 'living barns', a term we thought was perfect.

Before Lynbreck, we had primarily thought of the importance of woodlands as places for nature as well as a provider of firewood and just a nice place to be. We'd spent many years working with trees, cutting them down and planting them up, but the context was always about more woodland habitat for all the various bugs and beasties that relied on it. Now that we were becoming farmers, we began to learn how crucial trees were for our new business and farmed landscape. In fact, in some ways, they would become central to the very survival of our long-term plans and livelihoods.

Just a few months before our cattle arrived at Lynbreck, we had planted nearly a kilometre of hedging which snaked along our north-eastern roadside boundary. The little trees were a mixture of hawthorn, blackthorn, dog rose, hazel and holly, with the occasional cherry and rowan. Hedgerows are not a very common sight in our area but our neighbour George had managed to establish the most fantastic specimens around the edges of his fields using a similar mix of species and we always looked at them with envy. I had this dream of a network of trees and woodlands that would span the breadth of Lynbreck where you could walk from the south-east corner to the north-west tip at the top of our hill ground along a green superhighway and never leave the cover of trees. This connective corridor would widen and shrink, zigzagging across the croft and finishing in a full-blown forest that the wildlife and livestock of Lynbreck could navigate and forage.

Those early observations with our cattle taught us that the trees were just as important to them as they were for nature. We started to describe our new hedgerows and woodlands as creating giant tree hugs around our fields, providing future insurance that no matter where the high winds, rain and snow would come from, our cattle would always be able to choose to move into a place of shelter, crucial from a welfare point of view at least.

And, when they didn't have shelter when they needed it, this was also bad for business. On that hot July day, we noticed that in spite of being in the paddock for a number of hours, the cattle had touched very little of their fresh buffet of grasses and wildflowers. The heat had killed their appetite as they focused on trying to stay cool. When farmed animals stop eating for any length of time, this can affect their weight and overall condition, two factors which influence the final value of the beast when it comes to sales or slaughter. Although their welfare was what we thought of first, the lack of shelter might actually be losing us money.

I've had so many conversations with local farmers and crofters about the changing weather and climate over the last few decades, a topic I find fascinating coming from a country where every greeting is, 'Hello. How are you? Isn't the weather great/terrible/wet/windy?' We had continued that tradition as, when your life is so dictated by outside conditions, it becomes all you think about, your work day dictated ultimately by the weather. Our neighbour Fraser would tell us that summers have definitely been getting wetter, reminiscing about the long days of hay making by hand that have now turned into a race against time when you are fortunate enough to get a window of three days without any rain. And Hamish would talk about the winds, how they are becoming more frequent and more powerful than what he remembers from growing up in the area.

On the hot days, the animals would keep grazing if they could do so in the comfort of shade and, in the winter, our living woodland barns were their protection as they consolidated their energy for keeping warm, content to wait for their hay to be brought to them as the natural windbreak prevented the piles we laid out from disappearing like puffs of wispy smoke. They would happily tuck in, remaining in the shelter, before collapsing into the 'lumps' of cud-chewing contentment as the wild weather swirled ferociously above.

But in spite of all these benefits that we could see from our own experience, we were conscious of a definite cultural clash between trees and farming, where it seems to be either one or the other but never both. Today, when farmers apply for their annual basic payment subsidy, they often have to fully, or at least partially, exclude corners of fields where diverse pockets of scrub support mixed populations of insects, birds and mammals, all of which have a holistic benefit to the health of the food-producing landscape but for which the farmer is given no incentive to retain. Depending on the number of trees per hectare, groups of trees or woodlands can be deemed ineligible for direct payment support, even when the livestock benefit from the diverse mix of woodland forage and essential shelter. It did make us wonder if this system of incentivising one practice and giving penalties to another has helped to create the current scenario where hedges are ripped out, trees are cut down and nature is pushed back to the edges.

'Come on. Come on,' called Sandra as the entrance to the next day's paddock was opened. Without a care in the world, the cattle mooched through, heads straight down except for Flora who made a beeline for a single birch tree that stood at the bottom of the slope. Assuming she was looking for shade, we followed her down but we were wrong. We stopped, mesmerised as she stretched her neck and tilted her head back towards the tree branches above. Her huge tongue came out of her mouth, wrapping around the branch and stripping the fresh leafy foliage clean off. She was eating the tree leaves.

And we soon realised that it wasn't just leaves that our cattle were attracted to. They could sometimes be seen snacking on lichen that grew on the bark, their mouths delicately nibbling at the flakes and strands of white, grey and green that cling in abundance to their woody hosts. It was a trait that we had only ever associated with reindeer but our cattle were teaching us that they wanted it too. And, if they weren't eating from the trees, they were using them for a good old scratch on the trunks and lower branches, leaving behind tell-tale patches of smooth, orange bark and snapped stems with tufts of wiry red or black hair.

We started to change how we planned the cattle moves through the fields, trying to factor in access to trees for shelter and, if we could, for browsing as well. But what would we do come the wintertime? If the cattle were telling us that tree snacks were something they relished for extra nutrients and minerals, how could we factor that into their winter feeding when there were no leaves on the trees? If we were going to feed them grass hay in the winter, could we make and feed them a type of tree hay as well?

It seemed like a bit of a mad idea at the time but the more we thought about it, the more it made sense. When living and working in the south of England, we became acquainted with a brilliant and rather rambunctious character called Ted Green, the founder of a group called the Ancient Tree Forum. Ted had the most captivating personality and entertaining presentation style, which he used to relay an encyclopaedia's worth of information on everything to do with old trees. I had memories of him talking about tree hay, the traditional practice of harvesting branches in the summer to then dry for animal feed in the winter, but never came across anyone who was doing it.

It became our quest to find out more as a little research taught us just how valuable tree leaves can be, learning how animals could self-medicate by selecting the leaves of willow, which contain salicylic acid that provides pain relief, and how they would benefit from tannins found in tree leaves that have anti-parasitic properties and help to reduce internal worm burdens. This was a particular revelation considering that one of the most common medical treatments in livestock farming is worming, often administered as a preventative measure. For us, the turning point was listening to a talk by Dr Lindsay Whistance, an expert in animal nutrition and behaviour, who introduced us to an article produced by the Forestry Commission that said cattle could access up to 12 per cent of their annual diet through grazing trees and shrubs, a huge percentage that made us wonder if growing our own tree hay crop would be viable.

As always, a lack of capital was the biggest hurdle to overcome so we decided to approach the Woodland Trust to see if they might be able to help in any way. We drew up a proposal of planting, which included a bespoke edible hedgerow, a stretch of mixed native willow plants that would provide shelter throughout the year and snacking in the summertime. To our delight, they offered to fund the materials for the protective fencing and trees, somewhat intrigued themselves to see if this plan might actually work.

We identified a number of plots within our lower field where forage quality and diversity were low, not wanting to sacrifice the precious grazing we had on our upper in-bye fields. In total we planted five thousand trees, mostly willow species and a few alder, all of which would happily grow in the wet ground. The plan would be to cut some every year, gather branches into bundles and hang them to dry in our barn. They would then be fed out as a supplement to our bought-in pasture hay. We had no idea how much tree hay could be harvested, what percentage it would actually provide to our cattle's overall diet or how quickly the trees would grow, but it was an experiment that was certainly worth a try.

And while waiting for our new crops to grow, the plan was to harvest tree hay from what could be scavenged from in and around the homestead and wider croft woodlands. Although no one said anything negative, we're pretty sure it raised a few eyebrows in and amongst the local farming community. It was quite a contrast on a summer's day to see us with our handsaws, cutting and bundling fresh tree greens, stripped down to shorts and T-shirts because of the heat of manual labour. On the road above us, tractors with hay- and silage-making equipment went back and forth, causing the very air around us to vibrate with their sheer power as they drove past, rushing on their way from one job to the next during that precious dry weather window. Meanwhile we would cut, bundle and rest, the efficiency of our work not at all compromised by regular breaks in the sun in between.

Over the coming days, bales of hay and silage would start to appear in surrounding fields as our tree hay would be hung up in the barn, creating a light aroma of freshly cut green material that hung in the air for weeks afterwards. It was a job we so enjoyed, becoming a real highlight of the year where our bodies ached, dog tired at the end of the day, but our spirits felt satisfied and content, sleeping well after a long and active day in the sun.

'What is that? What do we do with that?' While we weren't fluent in cow speak, it was quite clear they were wondering what on earth we were bringing to them one cold winter day as we called them over to the fence line. Each bundle of now-dried leaves and branches were hung from the top line of barbed wire using cable ties. After a little bit of sniffing, nuzzling and occasional attempts at scratching, they started to tuck in, sometimes dislodging the bundles as they tugged, ripped and pulled.

In between the tree hay, we tied up bundles of dried nettles harvested in early summer when the plants were young and fresh. Nettles are a superfood for humans, having all sorts of purported health benefits from reducing inflammation to lowering blood sugar. They are also full of minerals such as calcium, iron, magnesium, phosphorus and potassium, all essential compounds for healthy life. Animals like our Highland cattle have an innate ability to tap into an ancient wisdom of selecting the plants that provide the very nutrition, energy and medicinal properties that they need and, with tree and nettle hay as supplements to their pasture hay, our goal was to make sure that they would have access to the broadest variety of plants that our land could offer. And they certainly loved the dried nettles, favouring them over everything else.

Diversity is what feeds nature and, on our journey to becoming farmers, we believed that it would be diversity in our farmed landscapes that could feed our animals and ultimately us. A healthy ecological system does not function in extremes, where hard lines are drawn between one habitat and another or where only one type of plant is grown or animal kept. We dreamed that our Lynbreck landscape would become one where mixed native woodlands would be connected by shelterbelts and hedgerows with herb-rich pastures full of wildflowers and grasses in between. Cattle would still graze, pigs still root and hens still scratch, integrated harmoniously into a multi-textured landscape, which they too would benefit from. Reflecting back on that early conversation with the farmer in the south of Scotland, I started to think, actually, we can eat trees. Trees help our cattle to stay protected and healthy and, combined with the forage they get from them, helps turn the beef into the highest-quality food for us. And the more we became practising farmers, the more our obsession grew about food, not just what we could produce for our community but what we could provide for ourselves.

CHAPTER 13

Food

I lined the roe deer up in my scope, the crosshairs searching for the kill zone behind the front leg where the heart is located as my own thumped so hard, the sound reverberating in my ears. After releasing the safety catch, I tensioned my finger lightly around the trigger, fighting the rising plume of panic and self-doubt that tried to overwhelm me. I hesitated, and hesitated and hesitated, becoming increasingly angry with myself that, in spite of the clear commands being given by my brain, my finger would simply not engage the trigger.

The voice in my head was literally screaming at me, 'JUST DO IT!' This moment wasn't just about killing an animal, it was about so much more – it felt as though the following action, to kill or not to kill, would define our whole future. How could we become farmers if I couldn't even kill a wild animal for our own meat? What if one day the law changed and small-scale producers like us would be allowed to kill livestock on site for sale? Could I do it? This animal was completely oblivious to my presence and, with one shot, could be dead in an instant. No stress, no panic, no anxiety, no pain.

But what if I didn't get the shot exactly where I wanted it? What if my aim was off? This was my first time shooting a real live animal and, in spite of the training and target practice I had done before, I was seriously doubting myself. I had heard stories of people injuring animals, not quite hitting them where they wanted, the animal running off to die a horrible, painful death in a bloody bed of heather.

I felt sick.

The rational side of my brain kicked in, talking to me smoothly.

Rational brain: 'OK. You are about to plant 17,400 trees. If these deer stay in here, they will eat all your trees. So, you need to do this. And you can ...'

Scared brain: 'What if I miss? I'm scared. I'm scared to do it. I'm even scared of the noise and feel of the rifle when the trigger is pulled. It's so powerful. And the deer. Oh, the little deer. It's just such a beautiful animal'

Rational brain: 'It is beautiful, and you can celebrate that, but this moment right now stands for everything you believe in. You can still respect the life that stands before you because this isn't about fun or sport, it's about so much more. It's about the new trees, the new habitat, your sustenance, your message and your journey into farming. It's the bigger picture. And you won't miss. You've got this. It's right there, standing quietly eating. It's so close and it can't see you.'

I knew in my gut what I had to do and began to shut down the voices in my head one by one. I took a deep breath, checked my aim and pulled the trigger.

We love food. We've always loved food. Early dates in our relationship often involved cooking together or going out for meals, and eating healthily was really important to us both. With most of our salaries going on rent and bills, our choices were mainly dictated by what we could afford, paying very little attention to where our fruit and veg came from, where eggs were just eggs and where organic was for posh people, way out of our price range. Our treat at the weekend would be smoked salmon with cream cheese on toasted bagels, one with freshly cracked black pepper, another sprinkled with a little bit of dill. It was our happy food. We had no real understanding of where the salmon came from, how that fish had lived or the environmental impact caused by farm-raised salmon. There was no connection made between the cream cheese and the life of the cow that had provided the milk for the cheese. And neither of us ever thought about the bagel and where the flour came from or how many chemicals had been sprayed on the wheat to help it grow, not even questioning the preservatives on the ingredients list. These were all things that were not at the forefront of our minds. We worked hard during the week and this was our weekend treat, which we so loved and cherished, washed down with at least one cafetière of nice coffee. The last thing we were looking for was a guilt trip, happy to merrily bob along in the sea of oblivion. And if we did feel guilty, why should we? If we couldn't afford organic chicken, did that mean we had to go without chicken? Neither of us wanted to go without chicken so we didn't, feeling defensive towards anyone who might suggest otherwise, as again, we worked hard and tasty food was one of the few things we could have as a treat.

I don't think either of us can pinpoint the exact moment when things started to fundamentally change. Our awareness of the realities of the food system was definitely there to an extent and something we occasionally talked about. But it wasn't until a few years into our relationship that we actually started to change our food-buying habits, when we could afford it. And, as our dream for a life closer to the land grew and grew, the conversation became increasingly dominated by producing our own food, to the point where it was pretty much all we talked about.

Before starting to raise and hunt our own meat at Lynbreck, our diet had become nearly exclusively vegetarian as we increasingly questioned where our meat came from, focusing on how the animals had been raised and the impact they'd had on the environment. We'd stopped buying meat from the supermarket and even asking the provenance from our local butcher wouldn't give us the reassurances we were looking for. While living in the south of Scotland, I had worked with a stalker contracted by my employer to reduce deer numbers in the areas where trees had been planted. We occasionally traded some beers for a few cuts of fresh venison, giving us our first foray into wild meat and we absolutely loved it. When we moved to Lynbreck I applied for my gun licence, trained as a deer stalker and bought my first rifle, a .243 calibre, which shot a bullet through a silencer with such ferocity that it became the most terrifyingly powerful thing I had ever owned.

I had been out stalking several times, but never really managed to get close to any deer as they often spotted me first from far away. As a prey species, their senses are geared towards picking up on predator movements, appearing to know instinctively when they are being hunted. It was primarily roe deer, an average-sized native species, that would visit the croft, living either solitary or in small family groups. When danger is afoot, they will often bark as they run off, raising their tail to flare up a bright patch of brilliant white that acts as a warning to any others in the area. I appeared to have a knack of unwittingly making my presence quite clear, as within seconds of making an appearance with my rifle, it seemed as if every deer in the area knew about it and a chorus of barks would echo out across the flats. Stalking requires tapping into what are usually forgotten and neglected natural instincts, taking us back to times of ancestral hunting where failure could mean hunger – and I definitely had some work to do on mine.

Shortly before the deer fence was completed on our hill-ground planting project, I spotted three deer in the enclosure and decided this might be a good opportunity to get my first successful stalk, hiding in an old wooden stalking shack in the gully and waiting until they descended into range. The young roe buck that I had just agonised over crumpled, his body falling a few metres down the steep sides of the gully and, within a few seconds, he lay lifeless on a heathery ledge. His companions fled the scene, darting straight out of the gap in the deer fence, leaving the area clear. I immediately called Sandra, who had heard the shot and was already on her way over as I started to scramble down into the bottom of the gully and up the other side in search of the body.

In those few minutes I wondered how I might feel when I saw the carcass of the first deer that I had ever killed. Would I be overcome with emotion and guilt? Would I be scared? I've killed many things intentionally in my life, mainly flies and midges, but this felt very different, even though in the eyes of nature, it was no greater or better a life than those smaller beasties I had squashed without a second thought.

The first task would be to gralloch the deer, the process of removing the viscera, which would be left in situ as a food source for any scavengers such as foxes, badgers, buzzards and eagles. I had never gralloched a deer before on my own, learning mainly from theoretical study, observing others and spending many hours on YouTube watching demonstrations. The closer I got, the more I wondered, would I actually remember what to do?

When I eventually found the body, I started to talk out loud, saying thanks for the life I had taken and promising to do my best with its preparation. It might sound silly but it felt reassuring and calming, shifting the focus away from me and my emotions as I had to make sure that this life was not in vain. By the time Sandra arrived, I had already made the incision into the skin, carefully opening up the belly to remove the organs, moving from one part of the process to the next, my mind going into autopilot. Once the task of gralloching had been completed, we each took a pair of legs and carried the buck back to the old croft house where we had rigged up a gambrel, a simple piece of kit which looks like a coat hanger with two prongs that stick out on either side, to hang the body and remove the skin. By this stage it was nearly dark and so, with head torches on, the head and lower legs were removed before we started to strip the warm hide from the cooling, slightly steaming flesh until the carcass was completely naked.

Rather than let it hang for a few days, we decided to butcher it there and then on a plastic trestle table, wanting to finish the job in one go. Anyone peering through the old cottage window on such a dark night would have had quite a fright as knives and blood dominated the bleak, dark interior. We bagged up everything, dropped some cuts round to neighbours as gifts and filled our freezer with the rest of the bounty, feeling content that we had done our best, with nothing going to waste.

Neither of us felt any guilt that night and, in all honesty, we felt immense pride. There was no glory or enjoyment from killing that animal and, if anything, it made us even more bewildered as to why anyone would choose to do such a thing for sport. But this was the kind of meat we could eat. The kind that had lived a natural life, eating food that was free of chemicals and drinking water that was clear of toxins and treatments. This meat would give us health and vitality from an animal that had lived the fullest of lives with the quickest of deaths, helping us to understand that if we were to farm, our animals would have to live as natural a life as possible and the challenge would be for us to ensure that this happened. From that point on until we started to produce from our own livestock, anything we hunted became our primary source of meat, and there was plenty on offer. Wild rabbit became our version of chicken as it has a similar taste, just slightly tougher, and which we would transform into a range of rabbit stews and curries, cooking large batches at a time to freeze, the meals feeding us for weeks on end.

Our first experience of eating meat from animals that had been raised under our guardianship was another significant step for both of us. It felt very different to eating wild venison or rabbit as inevitably we had built up a relationship with our animals through our daily interactions, knowing exactly their routines as we accompanied them throughout their lives with us. Tucking into Lynbreck pork from the first pigs that we sent away was actually quite exciting, the fullness of flavours taking us by complete surprise. Many people describe pork as a bland meat, something that needs a sauce or a seasoning to make it more palatable. But pork from a rare breed, free-range pig, one that has dined on a range of feed that includes natural vegetation, is completely different. And even the fat is different. I would usually have been quite picky about my meat, cutting any bits of fat out and feeling vaguely nauseous if I got a 'bouncy' bit, the kind which no matter how much you chew will just not break down in your mouth. Sandra, in contrast, is pretty unfazed by all aspects of meat, often hoovering up the rejected scraps on the side of my plate. But from our animals, I wanted it all. The meat, the fat, the bouncy bits. On a night when ribs were on the menu, we would merrily sit stripping meat from the bone like two hungry cavewomen, the kind of eating that you only do in private and certainly not what I would call 'first-date food'. Nothing from our animals would go to waste.

Growing, cooking and eating our own produce started to make up a large focus of our daily work, outside of setting up the new business side of things. Our newly built kitchen garden was already starting to produce in modest quantities after just a few months. My experience in growing food on a family scale was very limited, based primarily on my dad's pastime when, every year, he presented us with crisp sweetcorn and gluts of bright red and green tomatoes from a small greenhouse, complemented by sweet juicy strawberries that would tumble down the side of a blue barrel. Fortunately, Sandra came with more experience, having nurtured for a few years her own mini food forest on the banks of Lake Lucerne where, in a favourable central European climate and rich soils, plants would grow almost as you watched them.

But it was clear that there were many challenges to overcome in our quest to become as self-sufficient as possible, because in moving to Lynbreck we had taken on a very exposed landholding with a limited growing season due to our elevation and latitude. Sandra had made a couple of cold frames from an old wooden box with a hinged frame on top and a bit of plastic sheeting to catch the light and heat. She planted little trays of seedlings and started them off inside in relative comfort, but keeping a close eye as even in May we can get late frosts. By the end of the month, they were big enough to move outside into the main beds.

Before long, we soon realised that even down in the nearest town just five miles away, things could be a few weeks ahead of us. It would be mid June and most of our beds would still be brown, the little seedlings we had planted barely growing as if they were sitting in protest at being plucked out of their cosy nursery into this wild and unforgiving climate. With winds presenting the greatest challenge, we put a line of green mesh netting along the fence on the south and west sides of the kitchen garden, to give shelter from the worst of the prevailing gusts, replaced later with a living shelterbelt of native hedge plants. Along the north and the east, a dense wildflower strip would provide some seasonal protection, sown with a selection of seeds that our friends and family had given us for our wedding a few years previous. We were trying to create a living walled garden, one that would grow up in summer and die back in winter, but giving enough protection at the time of year we needed it most to create a more encouraging microclimate.

As summer was spent working in the kitchen garden, winter would involve planning for the kitchen garden. Sandra invested hours deciding what seeds she would buy and where they would be planted. Groups of plants were grown in families and every year the beds would be rotated, so that the same crop is never grown in the same bed it had been previously. Different plants have different nutrient requirements and can over time deplete those in the soil if planted in the same spot repeatedly, and we wanted to make sure that our soil fertility was gradually improved. Therefore in one year, a bed might be full of alliums that included leeks, onions, garlic and spring onions, whereas another might be full of legumes that included peas and beans. We planted brassicas that included kale, cabbages and brussels sprouts as winter crops, and for the summer we had lettuce, beetroot and radishes.

The beds were completely free of artificial chemicals, so it soon became essential to come up with more clever ways to protect our precious food crop from potential hungry wild animals and insects. The carrots were planted with our alliums to protect them from the dreaded carrot root fly, the powerful oniony and garlicy whiff throwing the fly off the scent of the sweet carrots in between. We used old bits of blue water pipe to make hoops over which insect netting was draped in mid summer. While early pollination of some of our plants was required, we needed to protect crops such as our brassicas from the butterfly known as the cabbage white, infamous for laying its eggs on the plant, which hatch into hungry caterpillars that literally devour the leaves. In that first summer, we made the mistake of not netting the brassicas and would spend hours picking off these little caterpillars and smushing them between our fingers, not the most joyful of jobs. And then there were the local blackbirds who simply loved the loose, worm-rich soils, often swooping down onto the bed, making a right old mess as they scattered the surface, dislodging newly planted seedlings and causing us to grumble in discontent, so more netting soon became used on other beds as well. We love nature and wanted the 150 acres of Lynbreck to be bursting at the seams with a whole range of birds, bugs and beasties, but our kitchen garden, a tiny 100-square metre plot, was sacrosanct and we began to defend it fiercely.

By the end of our first year, we managed to harvest a modest haul of lettuce, radishes, peas, carrots, beetroot, onions and potatoes and into the winter enjoyed hardier crops that included kale, neeps (turnips), Brussel sprouts and cabbage. Eating the produce that was grown from seed sown on our doorstep helped us to really connect with the core of Lynbreck, taking pleasure in every mouthful as we became initiated into what real, fresh food tastes like. A home-grown carrot is very different to a shop-bought carrot. Its natural flavours would sing a melody to our taste buds, its shape perfect in its imperfection when compared to the standard, almost freakishly identical, supermarket carrot. The radishes were the size of golf balls that were so juicy and spicy they gave a fiery heat to a cold side salad. And a pea eaten directly from its freshly picked pod would be heavenly sweet and, in spite of months of storage in the freezer, would retain that delicious taste as a reminder of our summer harvest, sustaining us during our winter hibernation.

The following year, our annual haul of veg became even greater, something we put down to experience as well as healthier soils. Our simple, somewhat higgledy-piggledy compost bays were starting to transform kitchen scraps, old hen bedding and grass clippings into beautiful, rich, crumbly compost that we used to dress the tops of each of the beds. And we discovered the wonders of nettle and docken tea. After harvesting the leaves of these weedy, prolific plants, which grew at Lynbreck in abundance and with such ease, we'd pop them into a bucket and cover with water. After just three weeks, the liquid was transformed into the most disgusting, foul-smelling green juice that whiffed of fresh poop. But when diluted and applied with a watering can, the plants guzzled up the goodness of this miraculous stinky water as our crops became healthier and more bountiful – fortunately the odious whiff did not come through in the flavour of the produce.

And just being and working in the kitchen garden became our happy place. It's hard to describe the atmosphere in there, why it feels different to other parts of the homestead, but it does. There is something about it that is so full of exciting life. While the beds would still be brown in June, the months of July to September would see everything explode into life as the many colours of the wildflower strip would come into their own and the vegetable plants would begin to produce in gluts. Maybe it's something in us that's very primal, that because we live in a world where shops are dominated by cheap, processed, mass-produced food, to be surrounded by such a buffet of healthy nutrition just makes us feel good. And it wasn't only us who felt that way but friends and visitors, whom we noticed would take visible joy from simply being in this little oasis. We would talk about taking trips to the Lynbreck Supermarket, browsing the different aisles and seeing what would take our fancy for dinner that night.

Before long, our growing skills improved so much that it soon became time to expand, wanting to push ourselves even closer to being as self-sufficient as possible from the land around us. One winter, we decided to enlarge the kitchen garden to the east, spending a week of hard slog digging out the grass turfs, which we transported into the field with a wheelbarrow to fill various holes and smooth over bumps. Down the middle we created a small gravel path with thin woodchip offshoots to either side and in between a whole range of soft fruit plants: raspberries, gooseberries, blackberries, redcurrants, blackcurrants, whitecurrants, strawberries and honeyberries. To cover the bare soil, we scattered the seeds of medicinal flowers such as chamomile and calendula, the tops of which we could dry and harvest for teas, with another wildflower strip around the edges.

A flat piece of ground to the north of the old croft house was perfect for upscaling our potato growing, a strip of land that we soon referred to as the tatty patch. We sowed a mixture of first earlies and second earlies, smaller varieties that are much like salad potatoes and need to be eaten quite quickly, and then main crop – much larger varieties that when harvested, can be stored in paper sacks to last the winter. We planted far more first and second earlies than we could possibly eat, so Sandra had the idea of dividing them into one-kilogramme portions and selling them roadside with our eggs, something that turned out to be a hugely popular venture as once word got out, locals would make the trip specifically.

In our second year of roadside tatty sales, we made enough money to cover the cost of all of our seeds for the kitchen garden, effectively rendering free any food that we grew. I estimated that we had reached a point where we were growing about 70 per cent of our annual vegetables so the trade-off for selling a few salad tatties to cover that was the kind of economics we liked. To afford the equivalent goods would cost a lot of money, money that we didn't have, but what we did have was time, which we could spend growing, nurturing and harvesting all that produce ourselves. And while there were days we grumbled about scavenging blackbirds or cursed the cabbage white, there really is nothing we wanted to do more or nowhere else we wanted to be. Working in the kitchen garden is the kind of time that you invest that money can't buy, which doesn't have a financial value and where time isn't money – time is life and nourishment. This was no longer a hobby, this was our lives.

But our lives also involved bills and no matter how much we would love to have traded our way out of those, we didn't imagine the local council or government office would have been overly amenable to us showing up at headquarters with a box of meat and vegetables in exchange for our annual council tax or income tax bills. Paying for our car and farm insurance was unlikely to work with an annual subscription of tatties. This was the 'real world' that everyone talks about and, while it was one we dipped in and out of in order to make the business side of our new venture viable, we had to be a part of it.

We'd dabbled in egg sales and managed to shift the first lot of Lynbreck pork, but none of those would pay the thousands of pounds we owed every year in insurances, taxes and utilities just to keep our heads above the water. And though we weren't motivated to just 'make money', we did feel as though we had questions to ask and points to prove about the relevance and viability of this way of living in a very busy, fast-paced, technologically driven twenty-first-century world. We had come to realise that farming with nature was not just about producing food while increasing biodiversity and soil health, it was something that would sit at the very core, the very heart of how we would run our business.

Nature is inherently diverse, and it is this diversity that gives it resilience. A business that was parallel in design, a diverse multi-enterprise venture, would help to give our own set-up resilience. In addition to our meat and egg enterprises, we had brought on a few hives of bees as pollinators but also as an additional stream of income from honey sales. If the bees didn't produce much honey to sell, we would still have pork available. Or if the hens stopped laying, we would still have beef to offer. And it would all be run on a solid foundation of collaboration. Nature is an intrinsically collaborative system where all elements work individually but for the benefit of everything else. The different elements of our business would be similarly connected, and our outward focus would be one of working with people and other businesses, and rejecting the notion of competition against others to succeed. Neither of us have any interest in competition other than pushing ourselves to do the best we can. When it comes to competing with others, we would rather turn our backs than engage in something we believe is ultimately futile, and we applied the same principles to how we would run our new business.

And when it came to setting targets for income and profit, the questions we wanted to ask ourselves were how much money do we need and how much money do we want? We've never deemed success in life as something measurable by the size of a monthly salary payment or numbers in a bank account. We knew what we had to earn to pay the bills but, above and beyond a bit extra for a rainy-day pot, we had no motivation to accumulate additional money for the sake of it. We would rather earn what we needed and focus the rest of our time and effort on life itself.

But, in spite of our resolve, there was still an undercurrent of worry, fear and self-doubt, the kind you feel when everyone else goes one way and you go the other. You know why you are going that way, but you can't help but wonder why no one else is? What do they know that you don't?

Either way, we'd reached the point of no return. With a croft full of animals and only a trickle of income from the last of our off-farm work, we had to make Lynbreck pay.

CHAPTER 14

Join the Club

'I quit my job.'

'What?'

'I quit my job. This morning.'

I had just collected Sandra from Inverness airport as she returned from a week at home with her mum and dad. This moment had been bubbling up for months and we both knew it. There had been endless back and forth conversations and any number of times when I had been on the brink of quitting but never quite managed it, the fear and worry of walking away from the monthly salary too great as we still needed money for all the bills that would pile in on us in what felt like a monthly avalanche.

'Oh well. I guess that's that then.'

And she was right. That was that. There was no anger or shouting or arguments. She just accepted that I had reached the time to quit and now we were really on our own. There was no point in going over the old ground that we had trodden so many times before. She didn't make me feel bad or guilty and, if anything, there was a slight vibration of excitement in the car as we drove home that day. The air felt a little clearer, almost like when we quit our jobs and moved to Scotland, when we first found Lynbreck, when our offer to buy it was accepted and we quit our jobs again. In the 'real world', it was another moment of utter madness, and in our world it was another time when, in our gut, this just felt right.

In the first few years at Lynbreck, our focus had been mainly on installing croft infrastructure – building a new barn, renovating the old byre, installing eight kilometres of new fencing, planting twenty-three thousand trees (a mix of new woodland, hedgerows and shelterbelts that would eventually become nearly thirty thousand), protecting an additional nine hectares of young woodland (that was naturally regenerating itself) and buying a whole suite of animals and kit that we would need to run our new farming business. Within thirty months Lynbreck had been transformed from a semi-derelict crofting unit into a fully kitted-out new agricultural business through an immense amount of hard work and by saving every penny earned, living within a tight budget, and accessing any rural grant funding that was suitable and available at that time. Lynbreck was now, finally, a place where producing food for sale more regularly could begin.

With twenty-five new hens arriving for the Eggmobile that summer to work as our pasture regenerators and soil builders, the pressure was on more than ever to find a sales outlet for all the eggs that would soon appear. With the best will in the world, we could not eat that many omelettes, cakes or meringues, and while our roadside honesty box was popular and great for surplus eggs, custom could be irregular and sales slowed down dramatically over winter. Until one day, I had a lightbulb moment.

'How about we start a subscription club?' I suggested to Sandra. 'We could invite locals to sign up for a weekly box, delivered to their door, which they pay for monthly and, in return for their commitment, give them a little discount on the eggs. We could call it Egg Club!' While I was particularly pleased with the idea, I was even more delighted with the name, dreaming it would become the coolest little club in town. Sandra agreed we should give it a go and so one evening we put out a post on social media inviting people to join. At the point of launch, I began to feel really nervous.

'Do you think anyone will sign up? I mean we think it's a great idea, but do you think anyone else will?'

'Look calm down, OK. Let's just give this a go and see what happens. It might just take a bit of time'.

Sandra has always been the steady ship between us as I bob around like a dinghy trying to navigate what can sometimes be a storm in a teacup. My impatience wants immediate results and when they don't happen, I always think the worst and I don't know why as panic can then start to set in. Sandra, on the other hand, is much more pragmatic and calm, keeping a grounded head as I huff and puff, usually completely unnecessarily. And this situation was no different from many that had gone before or would come after.

A number of our friends were immediate sign-ups as well as quite a few folks we didn't know, which was really exciting and evidence that word of our work was beginning to ripple through the local community. After two weeks, our subscriptions were all full and we were thrilled. We'd decided to charge £2 for a half-dozen box of eggs, a rate higher than other small-scale producers that sold roadside. It's a funny feeling when you are selling your produce for the first time and asking for money from it. You feel as though you just want to give it away and, for us, it was a strange and slightly awkward transition. And, with the price we set, it definitely made our eggs more expensive than many. But, because we were now trying to run things as a business, it was important to know our figures for what it would actually cost to produce the eggs – the organic feed, equipment such as feeders and drinkers, maintenance of their various buildings, delivery costs to town and even depreciation, a new but frightening hidden cost that we saw appear on our annual accounts. And what about our time? Normally when people go to work, they get paid for it. Should we and could we factor that into the figures? When we sat down to work through all of this, it was pretty frightening.

We knew that roadside sales of eggs at £1 or £1.50 a box were barely paying the costs it would take to keep the hens, something we had experienced directly as we sold our first honesty box eggs at £1.50. So why did people charge so little? For some, it's because keeping hens is a hobby to have a personal supply of fresh eggs, so getting rid of surplus by selling even at cost price would be the easiest route, often not wanting to turn the venture into a business enterprise. For others, including us at the start, it might be that awkwardness and uncomfortableness of asking for money, not wanting to charge too much, not knowing exactly where to set the price. But we knew with the figures in front of us that few people would be making any profit and were certainly not covering their own time, so why the reluctance to ask for a figure that truly reflected what the eggs cost to produce from hens that were well cared for and often with large areas to roam?

In shops, half a dozen eggs can start at less than £1, right up to £2.20 in our local supermarket, and more in others for the top organic range. While economies of scale can be one factor, we started to understand even more the realities and impact of the price of food. Food prices have been driven so low by retailers that people are now used to paying such prices rather than the true cost, resulting in the vast majority of farms today being propped up by annual government subsidies (many farms would go bankrupt overnight if the subsidies disappeared). A lot of producers who are selling direct often won't charge what it actually costs to produce because people will just not buy it. Unless and until customers understand why those eggs come at that price, they will question what extra they are getting for their money. Yet by charging a higher price, those on lower incomes are usually excluded from buying the best produce. With food poverty on the rise, many people can't even afford the basics to stay healthy, never mind make conscious food choices based on environmental impact and animal welfare. The growing disparity between more expensive healthy food for the privileged few and cheaper unhealthy food for those on lower incomes is one of the biggest injustices in society today.

The night before our first deliveries, I had a list of all the new members and where they lived so I could plan the most straightforward delivery route around the town, estimating that it would take about an hour to get round every house. Sandra had been preparing the boxes we would need throughout the week. With hens that lay blue, white, green and all shades of brown colours, she would carefully fill every box, checking each egg from that day's collection for cracks or blemishes and mixing the colours up so that customers would get a rainbow selection. It was the very antithesis of mass processing where Radio 4 would be playing in the background as she took her time to make them look as presentable as possible.

On the first delivery day Sandra stayed on the croft to work on chores as I spent a little extra time on my rounds getting to know our new customers. Upon my return, it became apparent that the estimation of an hour per full delivery round had been grossly underestimated. It wasn't that we had miscalculated the route, it was that the length of 'talking time' had not been taken into account, those few minutes when the customer sees you arriving and comes out for a friendly chat to converse on weekly happenings. As a result, deliveries were actually taking between two to three hours as I would have different catch-ups with different customers, not to mention the conversations with people I would bump into on the street or in local shops while picking up some shopping in between. Although this wasn't quite the 'quick and efficient' route we had planned, those two to three hours were some of the best invested and most enjoyable times of the week, getting to know our new members as they got to know us, building that bridge between their food and their farmer.

There was one customer in particular who lived in a first-floor flat, a lady we had never met before. I found the building and pressed the buzzer. 'Hello,' came the greeting through the crackly intercom. 'Egg Club,' came my reply. It was the first thing that came to mind to say, which came out in a sort of sing-song lilt. 'Oh, hello!' was the response, said in a way that you can hear someone smiling as they speak. I pushed the door open into the dark stairwell as she let me in and we met half way up, spending a few minutes chatting and getting to know one another. This is one of the lovely Egg Club traditions that has continued to this day.

Not only were we starting to get produce out weekly, we were seeing money coming into our bank account monthly. It was only a small amount but it was a step in the right direction to replacing our dwindling external income. To save a little more, I decided to try delivering eggs by bike. I could just about fit twenty-five or so boxes into my backpack and with a couple of paniers, another eight in each. The road down to Grantown, which fortunately is mostly downhill, was always a stressful ride as I cautiously navigated around every pothole and tiny lump and bump, acutely aware of my delicate cargo. I think Sandra thought I was a little mad, in some ways feeling just as anxious as she watched me push the bike up our stone track to the main road. But she just let me get on with it, knowing that if we could save any pennies at all, it would be worth a try.

One day, I had propped the bike on its stand as I walked over to chat with a customer. It was quite a windy day but I had become so distracted that I didn't take due care to properly check how steady it was. A few seconds later I heard a belly-sinking crash, accompanied by the musical ding of my bicycle bell. I knew immediately what had happened. The bike had toppled over, crushing eight boxes of eggs and leaving me short for that day's deliveries. I called Sandra.

'Do we have eight spare boxes of eggs and, if so, can you bring them down and meet me in town?'

'Why?'

The long, silent pause that followed my explanation said it all.

'OK, I'm on my way.'

Even though I knew she was annoyed, she didn't give me a hard time, knowing exactly what I had been trying to achieve, but we agreed from that point on that eggs on wheels would be much safer delivered on four rather than two.

The success of our Egg Club model was beginning to make us think. Our meat sales had started to pick up, with every release selling out a bit quicker than the last. We put a lot of time and effort into telling the story of the work the animals had done for us, with every customer getting a short information sheet to read that included details of where the animals had lived, the work they did and any particular observations we had made during their time with us. We'd made the decision only to sell locally, wanting to use our produce to feed those in our surrounding area. There was a huge temptation to offer postage and send parcels further afield to requests we'd had from big cities like Edinburgh, or even London, but we so valued the connection that our food could build with our community to their own landscape, and we wanted to nurture our market base there. Rather than cater for strangers hundreds of miles away, we encouraged them to find similar little farms in their area and give them the business.

And we were aware that our produce was certainly not cheap, again more closely reflecting the true value it cost to produce. After hours of trawling through websites, comparing prices to other similar farming set-ups, we were definitely at the upper end, giving ourselves an even greater challenge when it came to selling. And, just like we had learned with selling eggs, the enlightening and frightening thing was that even when charging higher prices, we were only just breaking even. When all the costs were added up, the return from our sales would match these and no more. We did manage to squeeze our own allocation of meat from that, a bonus which certainly was not taken for granted, but payment for our time? No chance. Depreciation of equipment? Dream on. Surplus profit for investment for new kit? Never gonna happen.

The food we kept back for ourselves was important. It was always part of our bigger vision of being self-sufficient from the land. But even that we didn't do very well in the beginning. The first time we offered pork for sale, we were so desperate to please new customers and make as much money as possible back, that we only kept a handful of packs of chops and sausages. Our focus had been so tightly fixed on earning an income and shifting meat as quickly as possible that the central reason for why we were living this life was forgotten – and that was to provide for and feed ourselves. It had been so easy to get caught up in the sales, to fall into that black hole of trying to make more income and being hesitant to say no that we forgot to quantify the value of keeping back stock for ourselves. In some ways it was frightening to see how quickly we had slipped into the money chase, vowing never to do so again. Meat as a source of food for us was far more valuable than any extra sales could ever justify.

And so we still weren't making enough money, certainly not enough to cover all the bills. If we took on more livestock, there would be more produce to sell but with that, more associated costs. And we were always thinking about the land, about carrying the number of animals in the short and long term that we felt would have a positive environmental impact. With current enterprises combined with our own workloads, there really wasn't much room for expansion. Nearby landowners had offered us grazing opportunities but, again, that was not the answer as with that came bills for renting the ground and extra work for which there was no time as we simply didn't have the capacity to travel to different sites every day. For us, it was time to get savvier, more efficient and think smarter rather than bigger, and we did have one idea brewing that might just help.

'Keep feeding it quicker.'

'I'm trying. It keeps getting blocked at the entrance.'

'Use something to push it down with.'

I was sweating buckets standing in our tiny cabin kitchen, scooping up handfuls of seasoned, minced venison that I was trying to push down the top hole of a little worktop mincer and sausage stuffer that my parents had bought us for Christmas. The motor would whirr and whizz as the mince was spat out the other end, gradually filling the sausage casings that were rolled up onto a long thin nozzle. I had to keep feeding the meat into the hungry machine at a consistent rate for it to come out the other side in a large sausage that Sandra would carefully craft into a spiral, making space to gather more as it came out.

We had decided to play around with a few basic sausage recipes, mixing salt and pepper, fresh herbs and spices, enjoying the wafts of the different aromas as they came together in what certainly smelled like culinary brilliance. There was no finesse or skill to what we were doing as sausage meat would end up in every crevice of our tiny cabin kitchen and the sausages themselves came out in all sorts of different shapes, sizes, textures and flavours. But it was quite clear from early on that Sandra had a natural patience and ability, mustered with a real respect for the butchery techniques that she had begun to research and learn.

Just a few hours later we tucked into an evening meal of home-grown roast potatoes, peas and our first homemade venison sausages, which tasted like nothing either of us had ever tried before. Being quite a lean meat, they were fairly dry and without any rusk, a common filler ingredient of breadcrumbs that is used in British sausages and tends to suck up and hold any excess fat. But the flavour was incredible. This was real food. Just good meat and seasonings mixed together, turning a simple banger into sausage brilliance.

Our own experimenting had got us thinking. Could we use our passion for food to turn our limited amount of produce from something that was already good into something incredible? Something completely unique? And something that by its very nature would be of higher value? It would involve better facilities, our own on-site butchery, a whole raft of new food safety rules and regulations to tackle, meetings with and visits by Environmental Health, a substantial amount of money to invest and, of course, learning how to butcher. While the many challenges to overcome were not to be underestimated, the idea excited us and, somehow, we would have to make it happen.

Sandra had found out about a day course over on the west coast of Scotland that would look at making charcuterie. The lady running the event claimed that by turning standard pork into charcuterie she could make £1,000 from one pig. With our current sales, we couldn't even make a third of that and her argument for 'adding value' was more than compelling, leading Sandra to buy books covering all sorts of artisan craft butchery, full of mouthwatering images of things like honey-cured hams, smoked bacon and artisan sausages. In the meantime, I had made a bit of progress on the funding side of things, coming across a charity called the Organic Research Centre who were offering farming units like ours the chance to apply for an interest-free loan to fund a diversification project. Our butchery idea fit the bill perfectly and I spent the next few months drafting another business plan, crunching figures and generally turning our latest idea into a paper reality. It was a process that we had to go through for the application but which we also absolutely had to get right for ourselves. This would mean taking on another loan, another financial burden and it had to be one that would pay for itself. But, just like other projects before, we were putting everything into this, believing it would happen and not even considering that it might not. Our energy was completely directed towards the outcome we wanted, our focus 100 per cent on the goal ahead.

And, for this project, every ounce of energy that we poured into it was needed. More figures were asked for, specific details on profit, even a site visit from a trustee to sound us out. And then, one day, the email arrived. I opened the attachment and, after reading the first sentence, went running out of the house.

'Sandra, we got it. We got the loan!'

'Oh god. Here we go again,' came the reply as we hugged, the next section of our adventure about to begin.

The following months went by in quite a stressful and intense blur. We made contact with our local Environmental Health Officer who talked us through the paperwork that would be needed. In between our growing number of chores, spare time was spent studying for food safety qualifications, writing risk assessments and house rules for our new butchery and, of course, overseeing the installation by our team of appointed builders.

Alongside our kitchen adventures, Sandra had been spending time with our farming neighbours Bill and Steph who had sold us our first Highland cattle. As two natural entrepreneurs, they had identified that in order to add the most financial value to their produce, it would have to be transformed into something more unusual, and so they began curing and smoking their Highland beef and Jacob lamb. They generously shared all of their learning with Sandra, taking her practically through every process, creating in the end a most delicious selection of smoked and sliced cold meats. And, as Sandra's natural flair continued to develop and grow, so did her inspiration and passion for bringing more of her own food heritage into the process, envisaging our new facility as a place to introduce Swiss culinary traditions into our Highland landscape.

During a trip to visit her parents, Sandra had found a one-day sausage making course that was running just thirty minutes or so from her family home and signed up. There she was introduced to the tutor Patrick Marxer, who originally trained and worked as a laboratory technician. Patrick had become increasingly frustrated with the disappointing quality of smoked salmon available in shops and so, after spending an entire night in the woods smoking fish over an open fire, he soon realised he could do a much better job himself. His passion quickly turned into his new profession, carefully crafting and adding value to foods produced to high ecological and animal welfare standards.

Patrick is passionate about sourcing natural ingredients for his products and has a knack for pioneering thinking when it comes to experimenting with different recipes and procedures, such as smoking and fermenting, combining all sorts of flavours. All this became more than evident during the day as he went on to explain how making sausages is a fantastic opportunity to bring together ingredients to create a real celebration of flavours, rather than just a way of using up all the lesser value cuts of meat, adding a few flavourings and bulking out the mass with rusk.

Most of the day was spent getting to know the technicalities and practical aspects before using hand-cranked stuffers to make Swiss bratwurst, Italian salsicce and smoked venison sausages, some of which were diligently tested at lunch time by the course attendees. One of the key messages Sandra took away with her that day was that there are practically no limits to what can be mixed in with meat and stuffed into casings. Adding herbs, spices, liquids and all sorts of other ingredients into that mixing bowl is a bit like alchemy, only that the goal is not to turn lead into gold or to achieve immortality, but to create the most wonderful taste experience, packaged as a sausage. That course was the building block which gave us the confidence needed to really delve into our butchery adventure.

'I can't remember where to start. I can't remember what to do.'

A half pig lay before Sandra on our shiny new stainless steel table in the now finished butchery. She stood there, staring blankly, dressed in butchery whites, hair net on and knife in hand. Trying desperately to stay calm, I could tell her anxiety levels were starting to rise.

Staying completely silent was my rather unhelpful contribution.

This was the culmination of hours of private study, YouTube bingeing, a generous butchery lesson from a local butcher friend, Jamie, and a hell of a lot of investment. We'd sent a few of our pigs off and decided to butcher them ourselves. Was this the point when we learned it was all a big mistake? That this time we had taken a step too far?

I chose to say nothing as my own levels of stress turned into an intense nausea, draining me of colour as my face turned the same shade as the assistant butchery whites I was wearing. I felt utterly useless but, thankfully, these are the kind of situations where Sandra comes into her own. Her stress turns into a fierce focus, a kind of determination where she just gets on with it.

After a few moments of staring at the carcass, her new shiny Swiss Victorinox knife (of course) pierced the skin and she was off. Within a couple of hours we had pork steaks, a fillet, a rack of ribs, a slab of belly and a loin for turning into bacon. We even had a leg roast tied up in butcher's string using a little knot trick that Jamie had taught her.

'You're incredible. You are absolutely incredible! Look at what you have done!'

I had the job of packaging and labelling, a task perfectly matched to my own personal butchery skills.

Those first few sessions in the butchery were long. We'd finish at 10pm and come in for something to eat and a glass of Prosecco. Neither of us know why Prosecco became our 'after butchery' drink. It's not something we regularly drink, preferring a beer or cider or a nice glass of red wine, but it just hit the spot after hours spent in the windowless space we came to refer to as the bunker. The cool, bubbly alcoholic sweetness seeped into our veins as we collapsed exhausted on the couch before getting up the next day and doing it all over again.

To keep up more regular customer contact and to give us the chance to be more creative with our produce, we started a new subscription club: the Little Mountain Meat Club, inspired by our location on the side of a little mountain. Once a month, fifty local members would receive a small parcel of our 'added value' meat range, delivered to their door and with the full story of the animal it came from. We made smoked streaky bacon, a classic hit, bratwurst using a recipe dating to 1512 and coming from a monastery in eastern Switzerland, whisky-smoked Highland beef and our own honey-cured back bacon. As confidence grew, so did our spirit of adventure. Inspired by one of our favourite Swiss sausages, we bought a massive chunk of organic cheddar from a nearby dairy, chopped it up into pieces and made a cheese sausage. We smoked pork shoulder, minced it up, mixed it into one of our classic sausages and called it a Smork sausage.

And, as much as we enjoyed the creativity, it seemed as though our new customers were also enjoying the experience. Some months our inbox would be flooded with emails. 'Wow that was amazing, do you have any more?' 'Incredible, best sausage I've ever had.' 'Can I sign up for a double subscription?' And other months, it would be silent. We started to get a feeling of what was popular, when we got it spot on, in contrast to when our latest experiment was a flop. After the first year, a staggering 94 per cent of customers signed up again for another year.

As well as bringing our customers closer to their food, even in just a small way, we were starting to see a return on our investment. The butchery paid for itself and a little extra, helping to close the gap on the financial shortfalls to turn our work into a viable business enterprise. And though that was essential, it was never our primary motivation. We loved that the animals we had cared for and raised, who had done so much good work for our land, had come home. Our responsibility for them when they were alive was as strong as the responsibility we felt for them in death when the real pressure was to do justice by them in the produce we would create. It wasn't about being morbid or macabre. This was the reality of what happens when you raise an animal that you are ultimately asking to give up its life for you, your sustenance and your business. These weren't our pets; these were farmed animals that deserved no less of a five-star treatment in death than in life. And now that is what we could do. Our butchery work became a celebration of their life, one that we shared with our customers, where people could not just experience what we created, but actually consume the very essence of what we were trying to share. That was a much more powerful, deeper connection than we could ever build with words and, looking back, this was us finally finding a place in our own version of what it meant to become farmers. A celebration that started with working with our land and animals and ended with the production of food and nourishing our community, all of which we played a practical and active part in.

And from what we could see, there were others out there who seemed to like this version as well.

CHAPTER 15

This Farming Life

'It's dropped. We're down to ninety centimetres and that's us barely using any water at all. There's no rain forecast for at least another week. So it means this is just going to continue to go down.'

The two of us were standing above the open manhole at the top of our well peering into the darkness as we searched for the pump and any signs of water. We'd tied a small rock onto a long piece of pink string, lowering it down until the rock hit the bottom and then pulling it back up. By measuring the distance between the rock and the top of the wet bit of the string, we'd get an idea of how many centimetres of water we had left.

The rock was on its way back up and from what we could already tell, the outcome was not going to be good.

After months of high temperatures and no rain, our worst fears had come true. We had just ninety centimetres of water left, barely enough to keep the pump afloat, which had started to tilt onto its side like a listing ship, a sure sign of doom. If the pump didn't have enough water, it would burn out and then the number of our problems would grow exponentially.

We had been emptying the well by three hundred litres a day to water thirsty hens, pigs and cattle without a second thought. Our fencers had just spent the evening power washing their digger after getting stuck in the mud down in our lower field. Sandra had been watering the kitchen garden daily. The list of use went on and on and not once, until now, had we thought to check our well, still conditioned by our previous lives where if you needed water, you turned the tap.

And the whole time, a camera crew was peering over our shoulders, filming everything for the television show we'd signed up to appear in earlier in the year.

'Well, it's your fault they're here.'

The stress of what was happening had caused a mini argument as I retreated back into the trenches of my defensive position. It was the kind of argument that never really gets resolved, where you go to bed without speaking and wake up in the morning to a frosty exchange, which gradually thaws throughout the day.

Nearly a year earlier, I came home from work and as soon as I got into the house Sandra said, 'There's a message from someone called Kate from the BBC. What's that about?' 'Oh yeah. I was going to tell you about that,' my face turning a shade of guilty scarlet.

One day in mid September, just eighteen months after we arrived at Lynbreck, a local farmer friend tagged us in a thread on social media. It had been posted by the BBC who were looking for new people to feature on their award-winning TV series *This Farming Life*, a fly-on-the-wall show that follows individuals and families for nine months as they share the realities of everyday living on a farm. It was a programme we had heard of but never watched as we didn't own a television set.

'I just thought I'd send them an email and see if anything might happen. I didn't think it would though.'

Thankfully, Sandra wasn't too annoyed or angry, but perhaps a little taken aback. We agreed that I should call Kate and see what she said.

Just a few days later, we found ourselves sitting side by side at our kitchen table wearing radio mics with a very smiley interviewer sitting opposite us, pointing a rather large, mildly intimidating TV camera in our direction. She asked us all sorts of questions about who we were and what our plans for the croft were before heading out for a wander across the fields and down to the pigs for some outside on-site filming. After about three hours, we were utterly exhausted.

And then everything fell quiet. We'd been told that filming would be starting in a few weeks' time and, upon hearing nothing, assumed that we had not been selected, an acceptance that felt surprisingly disappointing. Until one afternoon, a few days after New Year, my mobile rang.

'Hi. This is Jane. I'm the producer of *This Farming Life*. We'd love for you guys to be on the show. Are you up for it?' I'd been out fencing on the hill, so I dropped my tools and ran down the hillside, finding Sandra outside the cabin.

'*Farming Life*. They want us in it. And they're starting next week!' My breathlessness was in part due to the running mixed with a lot of excitement and a little bit of panic. Sandra stood in shock. Were we up for it? We certainly were.

Even before Jane had called, we'd talked through this situation. Why would we do it? Why did we want to be on the programme? Our passion was for growing our own food, the belief in the merits of truly farming with nature, regenerating our land, soils and community, our personal physical and mental health, all the while earning enough to pay the bills. We believed wholeheartedly that this way of working the land, this way of living, was not a step back to some romanticised ye olde days, it was just one example of how things might be done in the future. How small-scale units like ours could provide so much more than just food alone, helping to rebuild connections between people and the land. This chance to showcase it to the world was not an opportunity we could miss. There was just one problem with all of this. We were only at the start of our journey and they would be filming the pivotal year when either it would all come together – or it would all spectacularly fall apart.

The following week our small film crew arrived and, after some quick introductions over a cup of tea, we were kitted out with radio mics and it was time to get started. It felt a bit strange to begin with as we carried on with our daily lives, followed by a large camera on one side and a boom microphone on the other.

'Just ignore us. Don't even think that we're here. We don't want to get in your way.' Sentiments that we greatly appreciated but which practically were impossible. Every few weeks, the camera crew would return, spend a few days with us and then disappear until we had something interesting to film. We started to build a close relationship with them, sharing more of our real day-to-day lives than with many of our friends. They saw the intimate details of who we are and professional relationships became personal friendships, often sharing dinner of an evening after the cameras were switched off.

With the arrival of spring came more pigs followed by our first Highland cattle, introductions that brought with them a never-ending series of learning curves and firsts that proved to be ideal for the camera. They joined us on our first-ever visit to a livestock mart where we fancied buying a few sheep, a move I certainly never thought we'd make. I've never been a huge fan of them, a dislike I've not held back in sharing, but it was something that had given me a bit of friendly banter with our sheep-farming neighbour who, when he'd heard our plans, smirked gleefully.

We'd spied a couple of Jacobs that came with three lambs. Jacobs are claimed to be one of the oldest sheep breeds in the world and with their creamy white coats covered in black/brown splodges and a set of two or four horns, they were really quite pretty. We were drawn to them for their hardiness as well as the quality of their meat, the flavour of which was unbeatable, we'd been told. In the run up to the bidding round for the group that our sheep were in, a wait that meant staying until the end of the day, we felt so nervous. As the caller started the process, Sandra held up our bidding card, trying to be all cool and relaxed, but he didn't see it. She held it up again. He still didn't see it. Worrying we'd miss out on the sale, she held it up again as I dived to the right, waving my hand like a linesman trying to get the attention of the referee in a football match. Any attempts at looking cool and calm, especially on my behalf, failed as the whole auction chamber watched and the film crew caught another little bit of TV gold.

But it worked. That evening, we unloaded our small new flock and, in spite of the initial embarrassment, felt rather pleased with our purchase. The TV crew filmed us looking happy and content, a visage that turned to horror as, within a minute, we watched one of the lambs wriggle out of the fenced paddock into the field.

'Sandra?'

'Yes?' Came the very terse response as the presence of the camera all of a sudden became very intense and I giggled nervously.

By some sort of miracle, we managed to get the lamb back into the paddock but we realised that we would be spending the rest of our Saturday night fencing. Even the camera crew gave up as we walked up and down the fence line for hours, securing every little gap with wire.

By 9.45pm we were finally finished, tired and completely starving.

'Nachos for dinner?'

'Definitely.'

Tonight was not about health-conscious, local food. Tonight was about quick, easy comfort food washed down with a cold beer, and it was guilt free and unimaginably divine. I sat quietly thinking 'I knew it was a bad idea to get sheep'. And, unfortunately for them, their time at Lynbreck was short lived as we decided a year later to rehome the ewes with their lambs and take the one year olds to the abattoir.

But during the time they were here, we both became quite fond of our little flock, something which I certainly never expected to feel, especially characters like Nugget, an enthusiastic lamb that would often jump into the bag of hay in winter. They provided us with our first-ever lambing experience, a steep learning curve with some ups and downs, but an interesting time nonetheless. However, the economics of keeping such a small flock of sheep didn't stack up in their favour when set against the workload and extra management that had to be put in place. So we had to recognise that, at that point in time, sheep weren't right for Lynbreck. Naturally it threw up questions along the lines of, 'Have we failed? Are we not trying hard enough? Did we give up too soon?' But that began to subside as we felt the relief that materialised after making the decision, releasing the weight off our shoulders. We've never been afraid to try different things and make changes when they don't work. Farming and life in general are both fluid processes full of variables, and even though changes can be intimidating at first, they are often needed to take us closer to where we want to be.

It was true though. The meat was incredible. And Nugget went to a great new home with a local farmer.

The summer of 2018 was turning out to be an absolute scorcher and we lived in shorts and T-shirts as it was too hot even for the dreaded Highland midge to make an appearance. We showed the crew how we tried to work with our pigs in our fields, the practicalities of moving our new mobile pasture hens and grazing our cattle in ways that would help to regenerate our soils. They filmed us harvesting nettles, which we combined with meat from a wild rabbit and cooked up into a curry, and collecting our first hive of bees – a new journey into the world of beekeeping. But behind the idyllic vista of the sun-topped mountains, a real danger was starting to rear its head that would put everything we had been working so hard for in jeopardy.

Our well was nearly empty.

Living in the far north of Scotland, it is hard to imagine a lack of water but Lynbreck is located in the north-east, on the edge of an area referred to locally as the Moray Riviera, in part due to its low rainfall thanks to being in the shadow of the Cairngorms. As filming and the questions that came with it continued, the multiple implications of what this meant began to fill our minds as we accepted the likelihood that within a short time, our well would be completely dry.

Over the next few months, our daily chores and way of living took on rather a different pattern. The television crew followed us as we filled twenty-five-litre canisters of water from the River Spey on the edge of Grantown, a new daily routine to buy ourselves time to come up with a solution. I would pass the empty canister to Sandra who would stand barefoot in the river. When filled she would pass it back to me as I carried it, the water sloshing and rocking me from side to side, up a flight of steps, into the back of the car, and then repeat until all our containers were filled.

Throughout the dry period, there was a tiny spring that continued to run, pooling in a natural pond in our lower field. Using a small water pump, we managed to extract enough each day to bring to the cows in a bowser that was towed on the back of the quad bike. Our own water usage was cut down to the absolute minimum, relying on bottled water for drinking, river water to wash our hair and we took regular evening swims in a local loch to cool and soothe our bodies after long, hot, physical days of work.

It was a period of real worry and stress, accepting that this could be the future with climate change predictions suggesting an increasing regularity of warmer, drier summers to come. Water, in both scarcity and abundance, can be the limiting factor of life and the solution for us would be to capture and store water, so that in times of drought, there would always be options to draw on and in times of plenty, an opportunity to build up reserves.

Eventually, after a couple of months, a new well with a pump was installed by the tiny spring that continued to run, which gave us an independent source of water for our animals. We also added a giant 5,500-litre water butt to collect rainwater on the side of our recently built barn, and later additional water butts to capture rainwater from the buildings in the homestead. As the year progressed, our original well for the house began to recover, the groundwater levels slowly rising as a result of some very welcoming spells of rain.

The drought had cost us many thousands of pounds. Just as we had been watching the land become increasingly drier, exactly the same thing had been happening with our bank account. It was not an expense we had in anyway budgeted for or realistically could afford, but we had no choice. No water = no life.

The summer of 2018 opened our eyes and ears even more to this call to action that was coming from every corner of our land as we witnessed firsthand the impact of climate change. As our time with our camera crew was coming to an end, Sandra spent hours filming with the team, talking about our grasslands. She talked about the importance of building soil, how the trampled grass would break down and feed the very soil beneath, always regenerating life below the thick matt of vegetation. This was a fragment of a conversation that made it into the final episode and to the living rooms of many hundreds of thousands of people around the world. This was a message of farming with nature, farming with our team of animals, working hand in hand with both to share a positive story of producing food and life on the land. It wasn't about telling people how things had to be done differently, it was about showing one example of how farming could look, even though we were only at the very start of our journey.

Just a few months later, we had lined up a neighbour to feed the animals, allowing us to get away for the night on a rare joint trip to Glasgow. The BBC had paid for all the farmers featuring in the show to stay in a rather plush hotel as they treated us to a 'wrap party', an event that took place in the main BBC building and involved a generous supply of drinks, finger food and the viewing of the first episode. It was such a fun evening as we laughed, talked and let our hair down more than we had done in a few years. We had planned to have a long lie in, a leisurely breakfast where the goal was to eat as much of the free buffet as possible and then a slow drive home, with a coffee stop along the way. Instead we woke up to a text message:

'The cows have dislodged the handle on the water bowser and they are all out of water. What shall I do?'

'Well there goes the lie in.' And we hoovered up a few pieces of toast and jumped into the car heading for home. Sometimes it feels as though the croft doesn't ever want us to leave as every time we try to get away, something would happen to scupper the plans.

'It's starting. Eeeeeeek, here we go.'

Our friends had invited us over for a live viewing of episode two, the first where Lynbreck would make an appearance. I cried again, watching my reaction to taking our first pigs to slaughter, something that ended up making it into the series intro and repeated at the start of every episode, but overall we were quite pleased with how everything came across.

And, as the series continued, our worries about how we might appear slowly fell away, and we enjoyed the experience of watching our own lives unfold on the screen in a strange yet familiar parallel digital world. The skilled craftsmanship of the editorial team shone through, carefully weaving the subtle messages of regenerative farming and life on the land that we had wanted to impart to the wider world throughout the series. Our passion for nature and our animals was conspicuous, as was our determination to try new things and to share our first-time experiences. But, most of all, the programme showed just how much we laughed and smiled our way through what was one of the most challenging years of our lives, something we had completely forgotten.

The reaction from the public was overwhelmingly positive and we were interested to hear the thoughts of our peers, our fellow farmers and crofters, on who we were and what we were trying to do. While our approach to farming was quite different to the mainstream, we were still looking for some level of acceptance as we felt a growing affinity with, and an affection for, these people who faced many of the same challenges and hardships that were particular to this way of life and living. In spite of our smiling appearances and generally positive outlook, we can have feelings of vulnerability like anyone else and after showcasing our lives on the TV, these felt more acute than ever before.

Inevitably, a few comments filtered through that dismissed our work as 'not really farming', instead calling us 'lifestylers'. This is something that we know that other small-scale farmers face too – when some of our peers in the farming world look at the size of our landholdings and number of animals and brush our work aside as less important – and, to be honest, we don't really understand why. Lifestylers is a term that can have negative connotations and one which confused us as we were yet to find anything 'stylish' about this way of living and working. We had worked hard and made choices to get to this stage, navigating our lives to the point at which we found ourselves, always being thankful for the opportunities along the way. With such a positive story of farming, irrespective of size, we were baffled as to why any of our farming peers would want to dismiss it, considering the vast amount of negative press that farming receives today. Did they feel unsettled by our model of farming with nature? Or was it a wariness of a different way of doing things? Either way, the negative comments were few and far between and we decided to focus on the words of warmth and encouragement flooding in from around the world.

It was a baking hot day as I strolled along the River Thames, arriving at the base of Big Ben and trying my best not to get swept up in a swell of people, my navigation skills for this kind of situation now pretty much non-existent. A few hours before, I had been awakened by a gentle rocking and creaking as the Caledonian Sleeper, which I had boarded the night before in Aviemore, was on the final approach to London. My little bunkbed felt so cosy and a knock at my door brought a delivery of a steaming hot cup of coffee.

'This is all so exciting.'

In between filming, chores and other commitments, we had become members of a newly formed organisation called the Nature Friendly Farming Network, a collective of farmers who had joined together to call for better support and raise more awareness for the importance and need for farming with nature. I soon found myself on the steering group of the Scottish contingent, becoming increasingly involved in discussions centring around raising public awareness and actively canvassing politicians for their ear on how we could progress a more nature-friendly farming agenda.

The role involved invitations to attend events in the Houses of Parliament, spending the day dressed in smart clothes, hobnobbing with the government and industry elite. In some ways I enjoyed the buzz of the occasion in spectacular settings, but in others I was feeling more like a fish so far out of water my gills were gasping for oxygen. My smart clothes made me feel uncomfortable, a uniform I felt compelled to wear to fit into this different world of the political elite, but in which I felt self-conscious, stunting my identity and restricting my confidence to speak freely. I had no agenda other than playing my role in building a better and happier humanity, a purpose which in its purest and most honest form, seemed foolishly simple minded.

Nevertheless, our involvement with the network alongside an inspiring group of farmers continued and I was appointed vice chair of the Scottish steering group. More trips to both the London and Edinburgh parliaments followed and Lynbreck Croft became better known amongst political circles, hosting a string of visits from different Scottish MPs and organisations. Our work for nature was officially acknowledged as we were appointed Young Farmer Climate Change Champions by the Scottish Government and snippets of our story increasingly started to appear in regional farming magazines and campaigns.

Alongside our swift rise into the political and policy arena, accolades for our work began to appear, starting with the Cairngorms National Park Authority awarding Lynbreck Croft the Cairngorms Nature Farm Award. A few months later came the Scottish Crofting Federation award for the Best Crofting Newcomer, and shortly after the Newbie Network, an EU-funded organisation, announced us as the Best UK New Entrant Farm Business. In all of these, active farmers and crofters had selected us as the winners, extending the hand of acceptance and welcome from the community that we had so wanted.

And still the awards kept coming in. That summer Sandra and I were announced as the winners of the Best Farm Woodland Young Persons category by the Scotland's Finest Woods Awards and at the end of the year we were the winners of the Food and Farming category at the RSPB Nature of Scotland awards. Within just eighteen months we had won five awards, hosted endless groups, policy makers and politicians on the croft, written articles for magazines, recorded podcasts, been on national BBC radio and television, and attended countless events, meetings and conferences both in the UK and abroad. Lynbreck and our work was now well and truly on the map, something we were fiercely proud of and thankful for as the years of hard work and slog were really starting to bear fruit.

By this point, my days were spent answering emails, showing various dignitaries and interested parties around Lynbreck or spending time away from the croft altogether, invited to multiple events as our work was presented as a model of farming with nature, wilder farming, rewilding, agroecology, agroforestry, holistic management, crofting and smallholding. On the one hand, this was incredibly flattering as increasing numbers of people, some of which were high up in large organisations, were talking so positively about our work and we felt beyond grateful for their support. But while we seemed to fit the bill for multiple labels, all of which were relevant, we preferred to be defined by our strengthening individuality, which was quite simply: the Lynbreck way.

Meanwhile, Sandra was doing all the farm chores, trying to keep everything running smoothly, dealing with problems when they arose as well as trying to do the groundwork for the new enterprises we had starting up. Individually, our stress levels were on the increase again, both feeling overworked and tired, sensing that while we were excelling at sharing our work, we were failing at looking after ourselves. Little petty arguments began to creep in more regularly, tempers became shorter and quality time together was relegated to the bottom of the list. Our journey might be one of becoming farmers but it was primarily supposed to be life on the land, to be in nature, to grow and harvest our own food, and live at a slower, more relaxed pace.

Instead, things were moving fast and in some way our lives seemed to be returning to a similarity of that which we had made such a huge leap to leave behind. We found ourselves with incredible opportunities to stand alongside other farmers at the helm of political power, but started to doubt if we were the right ones to carry such an important responsibility. We were still so inexperienced, still only scraping a living from too many spinning plates. It felt as though we were just about holding everything in place but one wrong move and the whole thing could crumble. And, if we were completely honest, our expertise was not in policy and lobbying; there were so many others that had the ability and knowledge to fill that gap in which we simply didn't fit, and who would do a much better job in campaigning for a future for regenerative farming.

But while we didn't see ourselves as being the people to do that in a political sense, we felt our strengths lay with reconnecting people with nature and their food and helping as many others as we could to follow their path into nature-friendly farming and producing quality food for their communities. We started to realise that investing more time in our everyday lives at Lynbreck was going to enable us to do that much more effectively than anything else.

CHAPTER 16

Rooted

Many hours of our lives have been spent watching the view change from morning to night and through every season that comes to pass. It is a landscape that is dynamic and evolving, changing each second of the day and full of the sounds of life as the surrounding crescent of hills act like a natural amphitheatre, intensifying the call of every bird or bark of every deer.

In winter, all the details of the contours and features stand out so much more, poking out through the thick cloak of brilliant white snow that make the hills dazzle in the bright sun, turning a blueish hue when the clouds roll in. The short, intense burst of spring starts with a hint of green in our fields, before erupting into a vivid freshness of vegetation, all the while contrasted by the still brown and tired-looking moor, resting in its seasonal slumber. The air is full of a sweet scent of coconut coming from the explosion of bright yellow gorse flowers.

Summer sees the entire landscape turn a deep green, from our fields all the way up into the hills, culminating later on in the multiple shades of purple from the different varieties of flowering heathers. The homestead becomes busy with young families of blackbirds, pied wagtails and greenfinches. When autumn arrives, the annual cycle enters its final stage as the leaves of the birch become golden, the grass yellows, and the moor below is drained of colour once again as it returns to sleep, awakened only by the bright white scampering of a passing mountain hare or disturbed roe deer.

And while the landscape is big, the skyscape is a vast, powerful open space that can yield warmth from the sun or unleash the rains, winds, snow and hail during any season. Altering shades of light can make the valley beneath glow or highlight individual features with beams of sunlight that appear and disappear as the travelling clouds pass. As most of our weather systems get blown in from the south, we can often see what is coming before it arrives, allowing us precious minutes to finish up chores and race inside for cover. But in many cases, the downpours never reach us as we watch the clouds wrap around the end of the mountain range, releasing their load over neighbouring villages as we stay dry in the rain shadow of these mighty giants.

And, just like the landscape changes with the seasons, so, too, does the skyscape, which is most evident with the passing bird life. In the spring and autumn, migrating geese fly over in their characteristic V-formation, announcing their arrival and departure with their unmistakeable honking, communicating with one another as they work as a team on their long journeys south and north. In the summer, the audible pulse of snipe drumming in our wetlands vibrates through the evening air and in the winter, groups of fieldfare fill the trees around the homestead, swooping down into the fields in search of precious traces of food.

This changing view is the only television we have, our surroundings streaming a different live show every day and night, where passing wildlife from huge eagles soaring to tiny field voles scampering act as advertising breaks, reminding us of the wonder of nature all around. It was one of the reasons we fell in love with Lynbreck but, as life had become busier, our time spent observing, experiencing and enjoying it was diminishing. Our lives had become so saturated with projects and commitments that there was precious little space for anything else.

And that had to change.

I can remember times when we'd have family and friends visiting and, while I'd join them for dinner or a cuppa during the day, I'd be 'too busy' for anything else, running around the croft like a headless chicken while our actual flock stood relaxed, preening their feathers or snoozing by the woodshed.

And in some ways our relationship had become increasingly strained as we'd just forgotten how to be a couple, consumed by the pressures of life that swirled around us like leaves on a blustery autumnal day, uncontrolled and unpredictable. It was as if all of our energy was being sucked away as we shared with everyone else what we were doing and why, the demand on our time increasing.

However, we also acknowledged that we had steered our lives in this direction ourselves, in part unwittingly, not recognising the warning signs along the way, and in part purposefully by letting our enthusiasm grab every opportunity and challenge that presented itself. Now we could no longer avoid asking ourselves if this was really the life we had envisaged when we bought the croft. What changes would we have to make to start shaping our lives in such a way so as to create a more sustainable future on a personal level?

In the first instance, I decided to take a step back from our many external commitments: committees, steering groups and meetings, which in themselves had become a full-time position. We never set out to prove how doing it our way might be better than others. We had simply wanted to grow food and then to become the kind of farmers where our first job was to feed ourselves and our second job was to feed our community, and we felt our impact was indisputable there. Those who lived in our community, bought our produce, ate, talked, followed and interacted with us on social media, people directly within our circle of influence – these were the people who we felt we wanted to communicate with.

We had to press the reset button.

'Incredible. Just incredible. Look at that view!'

One of the upsides and the downsides of living at Lynbreck is every day we see the same view. While it's something that neither of us ever take for granted, it's hard not to be envious of people when they see it for the first time.

To help manage the growing numbers of requests for people to come and visit, we decided to offer a mixture of monthly public tours and 'anytime' private tours, charging a small amount for the former and a larger amount for the latter. It felt really strange and a bit awkward asking people for money to come and see us. We didn't want Lynbreck to have an entry fee where our knowledge and experience would be shared exclusively with those that could afford it. But at the same time we had reached a point where we ourselves could no longer afford the time, either for our business or personally. It was a way of helping us generate a bit more income and giving structure to our engagement time and we hoped that the information we shared and the experiences we provided would add so much more value to the lives of the people that came to see us.

Our tours would always start in the kitchen garden before extending out into the homestead and across the wider croft area. We'd share stories of growing food, talk with raw honesty about finances and introduce our animals while addressing head on harder-hitting subjects such as cattle and climate change or questioning if we should eat meat while standing next to the pigs. We'd stop for a break halfway through for a hot drink and slice of cake, getting to know the people who had travelled from all over the UK to see us. Our offer soon expanded to smaller tour companies, even hosting visits from groups as far as central Europe as we shared our story with an increasingly international audience.

On days when we were feeling tired and drained, the thought of having to host a tour in a few hours' time was not always something we particularly looked forward to, finding it difficult to muster up the energy. But every time, within a few minutes of people arriving, words flowed and by the end of the tour, there was so much positivity it felt as if we each had to fight an urge to have a big group hug. The unique energy of Lynbreck never failed to deliver until one day, it very nearly didn't.

'BOOOOOOO!'

The noise rippled through the auditorium as a few people reacted with awkward gasps and nervous chuckles. I was aware of a noise but just kept on talking, not even halfway through my presentation, as the audience quickly settled back into silence.

I had been invited to speak about Lynbreck at a conference on rewilding, the first of its kind in Scotland where over five hundred people filled a large amphitheatre and a range of speakers were invited to deliver a message of hope about the future of a wilder Scotland. By this point, we had substantially reduced our off-croft commitments but this was a unique opportunity to engage directly with a substantial number of people passionate about nature. I clutched my notes, which I had written out on index cards, my palms sweaty with anxiety, as I was called to the stage, trying not to trip up the steps that were awkwardly disguised in near complete darkness. The first of my prepared slides was loaded behind me and I began, acutely aware of the size of the audience and the importance of my message as over five hundred pairs of ears and eyes were all focused on me. I could not mess this up.

I wanted to highlight the role that farming could play in a rewilded landscape that includes people as a part of nature and which inevitably included the production of meat from the herbivore and omnivore teams. I started off by saying, 'We have to accept that it is OK to eat meat.' A comment that was not appreciated by a member of the audience, their discontent vocalised, as for the first time in my life, I was booed.

But, as I went on to elaborate, this wasn't just any meat. It was beef that was 100 per cent pasture- and tree-leaf-fed and pure rare-breed pork from pigs fed on a diet of organic feed, supplemented with what they grazed from the land. It was meat that came from animals that followed bespoke grazing plans to regenerate our soils and build biodiversity and were handled and treated with respect. It was meat that was full of nutritional goodness from a landholding where the net emissions from the croft as a whole were negative. A recent survey undertaken by the Farm Advisory Service and funded by the Scottish Government stated that at Lynbreck, over the period of a year, we sequestered over twelve times more CO_2E than we emitted. In essence, looking at our work and land holistically, we were capturing more carbon into our soils than we were releasing.

Other than that boo, the presentation was incredibly well received and, as I spent the lunchtime break walking round the various stalls of supporting organisations, a lady came up to me.

'I'm sorry. I was the one that booed you and I just wanted to say sorry.'

I didn't know what to say.

'I'm vegan and I didn't like that you said we have to accept it's OK to eat meat. But when I listened to the rest of your talk, I understood your side more and I just wanted to apologise.'

I'd obviously hit a nerve, the full impact of which I had not expected – and I don't think the lady in question had expected either.

'How did the talk go?'

I had managed to sneak a quick five minutes away from the crowds, hiding in a little wooded copse as I called Sandra with an update, not able to wait until that evening when I would arrive home.

'I was booed.'

'What?'

I think she was as surprised as I had been, both at the initial outburst and subsequent apology. It was a moment we have since reflected on quite a bit. We've always accepted that some people will not like or support what we do, especially those that have made different dietary choices for what are often very impassioned reasons. But, actually, this felt like a major breakthrough. Our role in giving this talk, as with everything, was never to try to convince people that ours was the 'right' way of farming or living, sometimes feeling uncomfortable when others, although often well intentioned, would elevate us to that status. It was about finding common ground, encouraging respectful dialogue, and building empathy and bridges. We all eat food and finding a supportive route to a better food system works to everyone's advantage.

We had reached a stage where both Egg Club and Meat Club were full and meat box releases sold out in a handful of minutes. We were starting to run our business as an actual business, one where income was higher than outgoing expenses and without being propped up by off-farm work. The tours were proving to be a success and we decided to write and launch our own course called 'How To Farm', the name inspired by Joel Salatin's book *You Can Farm*, which had made such a huge impression on us. The course was launched a few weeks before New Year and had filled within a couple of months. We couldn't believe our good fortune. With fewer commitments and a promising year ahead, we were starting to feel as though we'd finally got into our stride. And then the Covid-19 pandemic engulfed the world.

'The abattoir has closed. This is a problem.'

We had pigs that were ready to be taken off and, with new ones arriving shortly, the room for delay was minimal. As the world fell apart around us, so did most of our plans for that year. How could this be happening? Just as we said goodbye to the security of a job, just as we had really started to get into our stride, all of a sudden our projected income had taken a nose dive as anything involving people coming to Lynbreck was effectively cancelled.

Fortunately, the abattoir reopened in time for the pigs and while we seemed to spend painful hours refunding tours and course places, the demand for our produce went through the roof. From our meat and eggs alone, we could manage to cover most of our overheads but we only had a finite amount of produce. Thankfully, a number of kind course attendees agreed to carry over their booking to the following year and we qualified for a government Covid business support grant to make up the difference of lost earnings, both of which provided huge relief.

Those first couple of weeks when the world shut down felt strange. While we felt fortunate to live where we did, I think we both had moments of feeling down and I certainly had real times of worry. Until one day, while sitting outside having a cup of tea in the sun, I became acutely aware of just how quiet it was. Living next to a main road, which is also a busy tourist route, can mean weekends are like race day at Silverstone, but with the world in lockdown, no one was out. Here we were, Lynbreck all to ourselves for the foreseeable future, with nothing to do except run our little farming business and live. In some ways, this is exactly what we had been craving (not the global pandemic bit, of course), a slower pace of life with less external commitments, and now we had it in spades.

In between weekly deliveries, our time was spent working in the kitchen garden and across the wider croft. We would spend day after day seeing no one as we just got on with life. Earlier that year we had placed an order for a Polycrub, a super polytunnel-type structure designed on Shetland and built to withstand wind gusts of up to 120 miles per hour. Despite the sizable outlay, this was an investment into growing even more food and inching our way towards 100 per cent self-sufficiency in fruit and vegetables. The Polycrub arrived in a kit form later that summer and we spent a few weeks building it ourselves to save costs. While most of it was fairly straightforward to build, there were moments when swear words flowed like the River Spey in spate but, by autumn, the structure was up and watertight. I used to call it our food-growing cathedral as stepping inside felt like entering a spacious chamber where the very skeleton of the structure was exposed internally, making it feel very imposing. You couldn't quite hear an echo but I'm convinced it wasn't far off. The next spring our cathedral came alive with the growing song of tomatoes, peppers, sunflowers, strawberries, sweetcorn, courgettes, cucumbers, gherkins, squash, chillies, chickpeas and a whole chorus of herbs, the majority of which we felt fairly confident had never been grown at Lynbreck before.

And through all of this, finally we were making time for ourselves – and time for one another. We both started to read more, even meditate, anything to slow things down and really focus on enjoying the moment. We would barbeque as much as possible, not caring if it was a Tuesday night or a weekend, relishing the freezer full of meat we had to choose from, cooking up ribs, sausages, burgers or even steaks and accompanied by a freshly made salad from the kitchen garden. One of my favourite things is to stand next to the barbeque, smell the aroma from our homemade birch charcoal and watch the meat cook as I sip on a cold beer. After preparing the salad, Sandra would come and join me, often just sharing the time in silence as we inhaled every minute of those precious moments. It was perfect, as long as the rain and wind stayed at bay and the midges were busy elsewhere. On most barbeque evenings, at least one of those two weather elements or the flying biting machines would disturb us but, in some ways, that became part of the tradition we had to embrace, bundling our cooked food inside as we ate at the table by the window looking out over the Cairngorms. If wind, rain and midges is what we had to put up with for all of this, we'd happily make the trade.

With our enforced rest giving us renewed energy, we began to plan with real excitement for the future, our lives feeling more in harmony and balance than ever before. There was talk of planting dozens of fruit trees, shrubs and wildflowers in blocks in our fields, a way of increasing the food productivity of our land without removing much grazing for our cattle who would weave between. We looked at planting clumps of hazel, some for nuts and some for timber that we would harvest and cut on a rotation. And, for ourselves, the dream was a couple of goats or sheep to produce our own milk, cheese, yoghurt and butter. We talked of converting an old walled enclosure that was built into the hillside into a root cellar where we could store our preserved summer bounty of harvest as well as cheese and dried meats for our own consumption. All of this would need new infrastructure but the ideas alone were energising and, even though we accepted it might not happen now, we simply believed that it would happen in the future, targeting our collective focus at the outcomes we wanted.

When the world did open up again, our plan was to offer more courses, ones that shared the skills and knowledge that we had acquired and wanted to hand on to others to help them on their journey. Ideas ranged from a half-day kitchen garden introduction to charcoal making and foraging, all the way up to more of our How to Farm residential courses. It felt as though we were literally bursting at the seams with ideas, all of which brought back that real feeling of passion for what we were doing and how we were living. Ever since first arriving at the croft, we had always wanted Lynbreck to be a hub for people, and our evolving plans were now beginning to make that vision come to life.

But, in spite of our excitement for what the future had to hold, our life at Lynbreck still had its challenges, none more so than the sometimes difficult decisions we had to make, the most pressing of which right now was to do with Ronnie.

It was quite a straightforward decision in the end, the kind that we are quite good at making once we decide to go with our guts. One of the simplest ways to streamline our workload would be to change our model from breeding our own cattle at

Lynbreck to simply buying in young stock and finishing them for beef. It was turning out to be inefficient and impractical to try to breed from such a small number of cows and so the decision was made to sell our breeding females.

There was just one problem.

Flora and Flora we both knew were in calf but Ronnie, for the second time running, had come back from the bull empty. I used to joke that I thought she was so grumpy she probably fought the poor bull off but deep down I suspected it was not her temper that was getting in the way of mother nature. We couldn't in any good faith sell her on as breeding stock but how could we justify keeping her? Another mouth to feed with no financial return? We didn't have that kind of luxury. It was time for us to make a very difficult decision where there only seemed to be one solution. Ronnie would have to leave us that year for beef. In spite of very generous offers of rehoming, she was worth more to us financially as meat than people would be willing to pay. With the impact of the Covid pandemic hitting us hard and now all our external income gone, our budget was tighter than ever.

So we just had to suck it up and be farmers about this, right? Where there's livestock, there's deadstock.

We felt so torn. In terms of beef she would have been worth about £2,000, possibly a little more as she was naturally on the chunky side. This was a really significant amount of income for us. But this was Ronnie. Ronnie, our iconic black Highlander, the matriarch of the fold, our first ever cow. It was an incredibly hard decision to have to make.

Until one day, it wasn't hard anymore as we both agreed on what we had to do, making a decision that felt so right, we have never once doubted it or regretted it to this day.

On our journey to becoming farmers, we often put quite a lot of pressure on ourselves to do things 'the right way', to do things 'perfectly'. When we started our field grazing, we wanted to get it just right, beating ourselves up if the cows took a little too much that day, an error that was ours and not theirs. When the pigs got a little overexcited and dug too deep or too much, exposing more of our precious soil than we had intended, we were horrified, worrying if we'd broken that section of nature. And even in the butchery, when we made a mistake with a cut or simply stuck a produce label on at a wonky angle, we were frustrated with ourselves, never really cutting ourselves some slack. Looking back to those days, we were feeling the pressure as the intensity of the many sets of eyes watching us and ears listening to us grew and we increasingly felt the need to prove every action and justify every decision. We now realise there is no such thing as the 'right' or 'perfect' way, accepting all of our life contexts for what they are and simply doing the best we can for ourselves, those around us and nature at every turn; learning from our mistakes as a fundamentally important part of our journey and sharing those with open hearts.

And with our animal team, we had to learn not to get too attached, one of the hardest transitions of all, but Ronnie had got under our skin, as had many others like our resident hens Football, Rusty, Squeak, Charlie and Blue, ones we had kept back from rehoming and who would live out their retirement here. From speaking with other farmers, we learned that there are many like us, who not only have huge respect for their animals, but also have a soft spot for certain ones. And we learned and accepted that this is OK because we are human at the end of the day and these are real lives with real personalities that we are working with. Although many are destined to go for meat, we don't always have to cull out the ones that don't quite fit the intended bill. Sometimes, we just have to find them a new role.

Some of our retirement hens became characters that we would introduce on tours, giving some people the chance to touch a hen for the first time. We would tell the story of Squeak – how she loves to get into cars and regularly sneaks into the house. Squeak came to us from Donald McDonald, who had bought her as part of a bigger clutch of day-old chicks to raise. She could very easily have been allocated to a different batch that would have been sent to a factory farm, spending her life in at worst a cage and at best close confinement, never seeing the light of day, never able to stretch her wings, never able to jump, run and hunt. The story of Squeak was told to reconnect people with their eggs, to highlight the importance of buying high welfare.

And Ronnie? We had made the decision that she would stay, realising that she was more valuable to us now than ever before. With our decision to buy in new stock every year, we needed a constant lead figure who sat above everyone else and who would keep the fold in order and teach the new recruits the ropes. She was the matriarch, who would come when called with the rest of the fold following, and that was worth more to us than we could put a monetary figure on.

It's a comforting feeling to reflect and look forward on our Lynbreck journey by comparing it to the growth of a tree, in particular a species such as a majestic oak or a Caledonian Scots pine, the type of tree that grows for many centuries, where every crack or bark fissure is like a wrinkle on the smiling face of someone ancient, wise and kind.

In nature, when a tree seed is dropped, there is no certainty at all that it will germinate and become a tree and it must go through many processes before the first shoots break through the soil and into the sun. Our story so far feels as if it has travelled a similar path, where in those early days we scattered many seeds and, once we found our place at Lynbreck, there were many options for growth but only a few that would eventually succeed.

Our lives before Lynbreck were quite nomadic as we moved around, chasing new experiences in the search for our place in this world. It's only now, after five years, that we are starting to grow, having survived the early stages that have required us to dig deep within ourselves, become more resilient than ever before and reach for the light. It's a beautiful awareness to feel that we are now settled, growing and, most importantly to us, putting down those roots that we both so strongly desire to do.

The deepest root of all, which in a tree is called a tap root, forms first, delving vertically deep into the soil beneath, giving early stability. That is the one that each of us individually has started to grow, holding us in place at Lynbreck, while our other roots represent the connections in our lives that we now know are of the utmost importance as well.

Our social roots give us our place in a community as we develop new friendships and see the number of customers who want to buy our produce grow. It has helped to give us a sense of belonging as locals; we are 'the girls' and associated strongly with our work and lives at Lynbreck. And, although we are far away from old friends and family, Lynbreck has become a place of peace and reconnection for them, too.

Our environmental roots are helping us to become spiritually connected with the nature of Lynbreck. We began to realise that one of the reasons society has become so disconnected is because we have defined humanity as that which exists *outside* or *separate to* nature. The Cambridge Dictionary describes nature as: 'all the animals, plants, rocks, etc. in the world and all the features, forces, and processes that happen or exist independently of people, such as the weather, the sea, mountains, the production of young animals or plants, and growth.' Our world has become heavily influenced or affected by the dominant hand of humans as we are beginning to change the very climate we live in and the very genetics of the animals and plants we live alongside. This view of humanity as something separate to nature has fuelled a severing of our collective roots as many no longer respect the life we share our planet with, always taking and rarely giving, seeking to manage and control, draining resources and energy.

Our physical roots are those that connect us with the very essence that is Lynbreck through our food, that which we grow for ourselves and that which we produce for others as we have become farmers. Living and eating this way energises our bodies and minds as we spend more time than ever growing, harvesting, cooking and eating. For us, we accept this to be the core of what life is, to feed and nourish body, mind and soul, to know everything about our food.

And the most exciting thing of all? At the grassroots, there are many hundreds of thousands of small-scale producers, just like us, who are helping to turn the tide against the globalisation of food as a purely commodified item. As a networked collective, joined through UK organisations such as the Landworkers' Alliance and the international movement La Via Campesina, small-scale producers still grow the majority of the world's food and the recent revival of people wanting to return to the land, accompanied by a growing awareness in the public consciousness about the provenance of food, is revolutionary at its roots. These are the people who practice what can be referred to as 'conscious farming', enlightened farming', 'awareness farming' or even 'awakened farming'. A way of living and growing, harvesting and sharing, finding enjoyment and fulfilment, embracing the essence of the team of nature they work as part of.

We know how lucky we are. The reality is that there should be many more of us but, in spite of a growing will, that's currently not possible. We had to scavenge every penny we could find to make this happen. But there are many, unfairly, who can't even do that as cost and availability of land is the biggest barrier preventing more people like us who want to farm and live off the land. In a country where over half the land is owned by an estimated five hundred people, the need for land reform is pressing and urgent as the time to continue talking about it has long passed. If governments are actually serious about the climate emergency, protecting rural communities, supporting local food, promoting healthy diets, high animal welfare and protecting biodiversity, there is so much more that can be done, starting right now.

We see our future at Lynbreck as one that, like our view, is ever changing and dynamic. When people ask us 'would you do it all over again?', we say yes. When they ask 'what has your experience been like?', our answer can be summed up in one word: relentless. This way of life, way of living, just farming in general, is relentless. There are weeks when everything goes wrong, one thing after the next. It can take you to a point of mental and physical exhaustion where it feels as if every cell is rebelling to stop. But you can't stop. Because even if it is a blizzard, even if the shards of ice are hitting your face so hard it brings tears to your eyes, even if there are one-metre drifts to wade through and you are carrying large cannisters of water as everything is frozen, you have to keep going because the animals need you to. But the good times can also be relentless. When the sun just doesn't stop shining, the gluts of fresh produce from the kitchen garden keep flowing and the ongoing opportunities for evening beers and barbecues are too good to turn down.

As we realise our purpose, the shackles of the many pressures we put on ourselves in those first few years are starting to be released as we embrace more than ever our place in the world and believe that the legacy of our work will be evident in the nature that lives on when we pass. There are three certainties in life which ring true to this day: we are born of the soil, we live by the soil and we will return to the soil – the very word human coming from the word humus, which means earth or soil. We are part of a biological web of life that starts and ends beneath our feet, providing us with everything we need for a healthy life. It's this rootedness, this connection between human and nature, that we are consciously settling into as farmers. As we play our part in this journey, we wonder what life at Lynbreck will bring in the future.

About the Authors

Sandra Angers Blondin

Lynn Cassells and Sandra Baer met while working as rangers for the National Trust and soon realised that they shared a dream to live closer to the land. They bought Lynbreck Croft at the edge of the Cairngorms in the Scottish Highlands in March 2016 – 150 acres of pure Scottishness – with no experience in farming but a huge passion for nature and the outdoors. Now they raise their own animals, grow their own produce and are as self-sufficient as they can be, alongside producing food for their local community and hosting educational tours and running courses.

They were hailed as Best Crofting Newcomer in 2018, given the Food and Farming Award by RSPB Nature of Scotland Awards in 2019 and were nominated for Nature Champions of the Decade as part of RSPB's Nature of Scotland Awards 10th anniversary.

www.lynbreckcroft.co.uk
Instagram: @lynbreck_croft